# Palgrave Studies in Media and Environmental Communication

Drawing on both leading and emerging scholars of environmental communication, the Palgrave Studies in Media and Environmental Communication Series features books on the key roles of media and communication processes in relation to a broad range of global as well as national/local environmental issues, crises and disasters. Characteristic of the cross-disciplinary nature of environmental communication, the books showcase a broad variety of theories, methods and perspectives for the study of media and communication processes regarding the environment. Common to these is the endeavour to describe, analyse, understand and explain the centrality of media and communication processes to public and political action on the environment.

Mark Terry

# Speaking Youth to Power

## Influencing Climate Policy at the United Nations

palgrave
macmillan

Mark Terry
York University
Toronto, ON, Canada

ISSN 2634-6451 ISSN 2634-646X (electronic)
Palgrave Studies in Media and Environmental Communication
ISBN 978-3-031-14297-0 ISBN 978-3-031-14298-7 (eBook)
https://doi.org/10.1007/978-3-031-14298-7



Cover illustration: coldsnowstorm / Getty Images

This Palgrave Macmillan imprint is published by the registered company Springer Nature Switzerland AG.
The registered company address is: Gewerbestrasse 11, 6330 Cham, Switzerland

# Acknowledgements

My days as an environmentalist began before I had ever heard the word. My brother would take me on hikes and toboggan runs in Toronto's Don Valley when I was a child. I later would become a Boy Scout and upped my camping game. Travelling became a personal passion in my later years and I was able to combine it with my professional passion of filmmaking producing documentaries of exotic locations around the world in a time when they were called "Science & Nature" films. Eventually, the industry and the public started to call them "Environmental" films, and the now popular term, "Eco-Docs".

All this to say that many people throughout my own childhood youth as well as my adulthood youth have contributed significantly to my informed studies on climate change and youth activism.

I would like to acknowledge those supportive and influential individuals in this section in five distinct categories: My family, friends, academic colleagues, explorers, and UN colleagues.

*My Family* My patient, understanding, supportive, encouraging, and loving children: Herb and Mary Anne and their respective partners, Melissa and Ricardo; my grandchildren, Nico, Lina, and Allegra; my wise and compassionate brothers Hebé (Herb) and Chip (Bob) and their respective wives, Miriam and Sharon, as well as all their children and grandchildren, too numerous to name here.

*My Friends* Among my friends acknowledged here are those who provided moral support, guidance, encouragement, assurances, and who were always there for me when I needed a break. Your patience and understanding through this process are greatly appreciated and indirectly contributed to the completion of this book: Laura Bannon, Melanie Martyn, Michael Khashmanian, Janice Mennie, Stavros Stavrides, Carolyn Kelly, Blair Minnes, Michael Beighton, Gigi Stafne, Britta Schewe, Jackie Pearce, John McNamara, Timothy Hudson, Shirley Gossack, Sarwar Abdullah, Maeva Gauthier, Philip Jackson, Jon McSpadden, Tony Morrone, Vivian Guido, Emmanuel Marfo, Conor Fitzgerald, and the many Kichwa Peoples of Ecuador who welcomed me into their communities by bestowing upon me the Indigenous name of *Rasu Purik*, meaning "he who walks with snow and ice".

*My Academic Colleagues* Markus Reisenleitner; Susan Ingram, Martin Bunch, Sarah Flicker, Charles Hopkins, Katrin Kohl, Alice Hovorka, Rhonda Lenton, Jeff Grishow, Russell Kilbourn, Sandra "Kushi" Sanchez Gordon, Arawi Ruiz, James Orbinski, Theresa Dinh, Shana Yael Shubs, the Dahdaleh Institute for Global Health Research, the Royal Society of Canada, York University, Wilfrid Laurier University, the Quechua Academy, and the Escuela Politecnica Nacional University. Special thanks to Geetha Selvapandiyan, Arunaa Devi, Shinko Mimura, and Lauriane Piette of Springer Nature and Palgrave Macmillan for their patience, assistance, and guidance through this book's publishing process.

*My Explorer Colleagues* John Geiger and Joseph Frey of the Royal Canadian Geographical Society; George Kourounis, Maeva Gauthier, and Ginny Michaux of The Explorers Club; Lorie Karnath of The Explorers Museum; Tim Lavery of the World Explorers Bureau; and the anonymous Member of Parliament from British Columbia who nominated me for the Queen's Diamond Jubilee Medal.

*My United Nations Colleagues* Nick Nuttall, Catherine Beltrandi, Elisabeth Guilbaud-Cox, Fanina Kodre, Achim Steiner, Angelica Shamerina, Tina Cobb, Laura Coates, Melita Kolundzic Stabile, Adriana Valenzuela Jimenez, Fleur Newman of the United Nations Climate Change secretariat; Cassie Flynn of the United Nations Development Programme; Charles Hopkins and Katrin Kohl of UNESCO; Nick Turner of Earthbeat; and all the members of the Canadian delegation to the United Nations Framework Convention on Climate Change.

# Contents

# List of Figures

# Introduction

There exists much scholarship on youth climate activism largely centred on the phenomenon and efficacy of the global youth climate marches known as the Fridays for Future protests as well as its impassioned leader, Greta Thunberg, and the representational politics of her eco-celebrity (Murphy 2021). Media outlets, academia, and the general public are keeping a close eye on the measurable impacts of such activism. Few question that these well-organized global protests conducted by youth have raised the level of awareness of the climate crisis among the world's population. Many also agree that they have increased pressure on all levels of government. However, this final step has yet to be realized to the degree that is necessary to arrest "the increase in the global average temperature to well below 2 C above pre-industrial levels and (to pursue) efforts to limit the temperature increase to 1.5 C above pre-industrial levels" ("Article 2,1", 3). Systemic change in the way the world engages with fossil fuels and other greenhouse gas-emitting practices and industries is often demanded, but is slow to evolve and be realized.

And when we get to this stage of environmental policy creation at the level of the United Nations, we see that the global community of youth is not present. Aside from participating in large numbers for protests, they do not seem to participate in large numbers in the global policy process regarding climate change, the cause they so passionately embrace today. This book will examine the various reasons for this such as their perception by power of being inexperienced in political processes, being formally uneducated in climate science, being seen historically as destructively anarchistic in nature, and generally, being too young to provide any meaningful contribution. The investigation of this book posits that youth should be regarded as a global community marginalized by ageism and should be acknowledged in the same manner other marginalized global communities like Indigenous peoples are recognized in participatory politics. In particular, this book will look at how the United Nations is experimenting with programs and methods that begin to acknowledge this global community status and to take this group's unique perspectives and cultures into consideration during its negotiation sessions at the annual climate summits known more commonly as the COP conferences.

In an attempt to be comprehensive in this investigation, this book is divided into two sections of inquiry related to methods of engagement youth are currently using to achieve their goals with those in power: the *Groundswell Approach* and the *Direct Approach*. After examining the historical and theoretical aspects of each approach through existing scholarship, a framework for productive policy participation between youth and power is proposed and applied to current methods of engagement used by the United Nations Framework Convention on Climate Change to involve youth as a global community member in its annual climate policy sessions.

Chapter 1 looks at the foundational origins of youth activism dating back to 1815 through to the 1990s. The author acknowledges that there are many more youth movements during this period, but has selected these examples in particular as they best represent initial perspectives and methodologies of youth activism that have informed and inspired the behaviours and practices of today's youth climate activist.

The first example dates back to 1815 when the *Burschenschaft Movement* began in German universities. Informed by the work of historians L. F. Zwicker and Priscilla Robertson, we examine how European youth in Germany rebelled against the values of their adult counterparts in ways that are not unfamiliar to those who remember the hippies of the 1960s, distinguishing themselves, for instance, with unconventional attire for the time. These early youth activists sought for a dismantling of class structure, but not without disorder. The five university fraternities that originally formed the movement insisted their members continue with political education and gravitated to a "political stance that was Protestant, liberal, and German nationalist" (Zwicker, 399), a unique community of young people that represented "a new brand of liberalism" (Robertson, 368–9).

Elsewhere in Europe, the 1830s show us our first political youth movement, *Young Italy*.

The writings of the founder of the movement, Giuseppe Mazzini, a revolutionary of the time, and Lorraine Murray paint a picture of youth impressed by a charismatic leader who in a century and a half before the Internet grew his number of followers from 40 in 1831 to 50,000 in only two years, impressive even by today's social media standards. With a blend of Church and state in his rhetoric, Mazzini demanded his followers to take an oath of allegiance promising membership devotion to a cause that aimed to "constitute Italy one and free, independent, republican nation". While this sounded good to Italian youth in his speeches and on paper, its practice left something to be desired. The group's attempts at government insurrection in six Italian cities all met with failure. As inspiring a leader as Mazzini might have been, he could not mobilize his army of youth to violence and anarchy.

Seventy years later, another youth movement emerged in Germany this time not formed by students attending university, but by a teacher and nature lover named Karl Fischer. In 1901, he selected youth for camping expeditions outside of the urban centres and perhaps provided the model for the Boy Scouts seven years later. Here the work of historians Gabrielle Nagy, David I. Macleod, Oliver Coburn, John Alexander Williams, and Jon Savage are sourced to describe the *Wandervögel* ("wandering bird") *Movement*. A sense of community is again prevalent in these youth groups as is their desire to be separate and distinct from adult communities. This time, however, we see our fist keen interest in the environment and "field trips" as camping defined this movement. Organized by Fischer, parents were fine with these expeditions until girls began to become a part of the movement. Splinter groups soon formed that continued the field trips without adult supervision. One of the first European youth movements to expand outside the borders of its country of origin, the popularity of Wandervögel spread to Switzerland. Here, the young environmentalists evolved the movement and "created a commune that practiced vegetarianism, nature cures, and free love", another precursor to the hippie movement of the 1960s.

In 1918 in Argentina a new kind of youth movement was born out of the universities in Buenos Aires, San Miguel, and Córdoba. Research conducted by Natalia Milanesio, Rodrigo Arocena, and Judith Stutz informs us that the *University Reform Movement* was started by students rebelling against the structure of the Catholic-run school system that had been in place for hundreds of years in this South American country. They accused their professors of being "authoritarian, inefficient, clerically oriented, and obscurantist" (Milanesio, 505) and were ultimately successful at reforming the post-secondary education system through a series of strikes, rallies, and petitions, an early indicator of the "School Strikes for Climate" popularized by Greta Thunberg 100 years later. This youth movement is significant in examining the youth climate movement today as it represents an educated group of youth (university students) applying their knowledge to recognize a flawed system (their own academic administration) and taking disruptive action (school strikes) to achieve positive change (university reform). Another interesting parallel to the youth climate movement of today is the international appeal of the cause. The success of the University Reform Movement in Argentina found followers in Peru, Chile, and Mexico between 1919 and 1921, all eager to achieve the same success in their own countries.

Forty years later in North America we begin to see one of the first transcontinental, if not global, youth movements: the Hippies and Yippies of the 1960s. Columns, articles, reports, and scholarship from Edward Kern, Michael Mandelbaum, Daniel Walker, Lee Jamieson, Sherri Cavan, Dan Rather, David Taylor, Sam Morris, and Abbie Hoffman paint a picture of a youth movement that grew in unintended directions and size. Beginning much like the youth of the Burschenschaft Movement with non-conformist clothing and the move-to-nature, drop-out ideology of the Wandervögel Movement, the hippies soon posed a threat to the status quo and became political in spite of itself. The subsequent splinter group, the Yippies, picked up the political mantle with the anarchistic and revolutionary passion the hippies were not prepared to embrace. The violence so eschewed by the hippies' mantra of "make love not war" was introduced in full force with the "police riots" of 1968 at the Democratic National Convention in Chicago.

Significant in the evolution of youth climate activism in this era was an early interest in natural environments as opposed to urban environments and the political nature that the movement soon became. Also noteworthy is how these movements communicated to adults using artistic expressions (music, flash mobs, theatre). Making its appearance on the world stage was the television and its first international broadcast of a disgruntled community of youth attempting to disrupt the status quo and introduce social change. The hippie/yippie movements would not have grown to such global numbers without the aid of this electronic communications system. This will later serve as a model for digital communication through the Internet for a new generation of youth looking to dismantle governments that not only tolerate but support systems that enable damage to the planet and the growth of climate change.

The final era this chapter examines are the decades that immediately followed: the 1970s and 1980s from a global perspective. With a healthy distrust of authority borne from the 1960s' altercations with police, the establishment, and "the man" in general terms, youth became more sceptical of both government and industry worldwide. Here we look at the phenomenon of Rachel Carson and her influential book *Silent Spring*. Described as a "fable", this work of fiction is based on facts related to a poisonous chemical found in pesticides at the time known as DDT (dichloro-diphenyl-trichloroethane). This artistic expression had a profound influence on the Kennedy administration in the USA resulting in the banning of the substance. Dubbed "The Gentle Subversive" and an "improbable revolutionary" by Mark Hamilton Lytle, American youth saw first-hand that you did not need to be a scientist, a politician, or even a trained journalist to be able to cause progressive social change in peaceful and non-violent ways. Claire Patterson, Barbara Scott Murdoch, and William Kovarik shed light on another environmental concern this time caused by leaded gasoline. In 1970, another victory over industry was achieved when the newly formed Environmental Protection Agency ordered car manufacturers to begin building engines to run on unleaded gasoline. Another careless use of environmentally dangerous chemicals and gases by business and industry emerged with the discovery of the ozone hole by Jonathan Shanklin of the British Antarctic Survey in 1985. However, unlike the other two examples of the use of DDT and leaded gasoline, industry did not resist to the same degree the banning of the substances they used to great extent in the manufacturing of aerosol spray cans. In fact, for the first time ever, the United Nations was instrumental in drafting an international agreement to ban the substances that was eventually signed by every country in the world. The Montreal Protocol gave a new generation of hope to youth that science, government, and business can come together in the face of a global environmental crisis to solve the problem, a hope that proves to inspire and fuel the youth climate movement in the years to come.

Chapter 2 examines the Groundswell Approach to youth climate leadership. There has seldom been a cause with so many high-profile leaders among its youth. This chapter will look at the inimitable Greta Thunberg, the current face of global climate youth activism, and those lieutenants in her ranks who attempt to imitate her reach, strength, and passion, but serve effectively in disseminating her calls for action and expanding the reach and recruitment of the cause's global ranks. For a foundational understanding of youth's role in environmental causes, we take a page from Buckminster Fuller's apt metaphor of "Spaceship Earth", his ecological concept that humans are passengers on an intergalactic vessel and that we on this ship need to work together to keep it in proper working order, otherwise if damaged, the lives of all who travel in it are lost. This 1969 reference leads well into an early environmental youth group founded in 1990 called Eco-Kids, a youth movement that *TIME Magazine* called a "new generation of conservation-conscious, environmentally active schoolchildren" (Elmer-Dewitt, 51).

The phenomenon of Thunberg's appeal, power, celebrity, and influence will be seen through the lenses of research published by Jennifer Wright, Jonathan Watts, and Dorothy Lepkowska. To better understand her leadership appeal we apply the model developed by James MacGregor Burns in his Transformational Leadership Theory and its expanded version proposed by Bernard M. Bass. In applying the four elements of a transformational leader—*Idealized Influence*, *Inspirational Motivation*, *Intellectual Stimulation*, and *Individualized Consideration*—an additional context is provided to understand the success of Thunberg and her fellow youth climate leaders. With this theory applied we add the dimension of digital communications to understand how youth today expand their reach and their messaging and thereby magnify their presence and appeal to celebrity status. This chapter also briefly explores the power of such status and how relatively easily it is achieved in today's digital age.

A 2020 study conducted in Canada by MacEwan University (Edmonton) and Université du Québec (Trois-Rivières) reveals some enlightening data about the use of social media in organizing and broadcasting the Fridays for Future protests, most notably, that very little social media posts have anything to do with participant recruitment (only 4.8% on Twitter, for example). The impressive numbers of attendees and the global coordination speak to the success of the protests as an event, but critics such as Jens Marquardt are still waiting for tangible results and measurable impact.

The chapter concludes with case studies of youth climate activists who have not only subscribed to the Groundswell Approach and participated in climate strikes, but who also subscribed to the Direct Approach addressing the United Nations and its policymakers face-to-face.

Chapter 3 explores the methodologies being used by youth today as they apply the Groundswell Approach to their climate activism. Never before has the global community of youth been so connected with communications taking place across the world instantly. For this chapter we examine the extensive research made by renowned digital media scholar Henry Jenkins first theoretically with his book *Spreadable Media* which provides keen insight on how youth are using social networks to advance their cause and recruit followers. We will then examine his more focused study on youth activism, *By Any Media Necessary*. By learning how youth engage with each other digitally in a participatory culture, we will learn how they are applying these practices to establish a similar relationship with the policymaker in a form of participatory politics. The research of sociologists Jacquelien van Stekelenburg and Bert Kandermans contributes to this discussion. This chapter will also review the latest statistics of use among the Internet's most popular social media platforms revealing some surprising results. Contextualizing this data for young climate activists is provided in a study conducted by Emily Wielk and Alecea Standlee in their research paper *Fighting for Their Future: An Exploratory Study of Online Community Building in the Youth Climate Change Movement*. Further social media analysis is provided by Christoph Herrmann, Sebastian Rhein, and Isabelle Dorsch whose study on networked social movements proposes three distinct identities among its users: individual, collective, and public. All three contribute to the cause, but in different ways. Understanding this allows for better targeted campaigns and more successful results.

Political anthropologists Yarimar Bonilla and Jonathan Rosa shed light on the use and power of the hashtag (#) in supplemental information binning and specific communications. Their study, *# Ferguson: Digital Protest, Hashtag Ethnography, and the Racial Politics of Social Media in the United States*, presents the hashtag as a form of indexing system for micro details of macro events. It is also used as an entry to the climate discussion for the "lazy" climate activist by adding a hashtag to a comment or a picture defining the term "slactktivism" as coined by Ellen Simpson in 2018. While this might seem "the least they can do", it actually supports the Groundswell Approach in large numbers bolstering the global presence of youth climate activism online.

The chapter also includes the perspective of youth climate activist Jamie Margolin, a 21-year-old (as of 2022) American climate justice activist and co-executive director of *Zero Hour*, a "youth-led organization demanding climate action".[1] We include her work *Youth to Power: Your Voice and How to Use It* as it serves as a "toolbox" for youth climate activism, according to Greta Thunberg, and speaks to the social media methodologies and options available.

Finally, this chapter identifies a couple of terms that are used throughout the rest of the book: the *Citizen Scientist* and the *Citizen Journalist.* These titles reflect untrained people working in the fields of research and reporting usually under the guidance and direction of professionals in the field. Their contribution to research, data, and documentation manifests a new facet of the Groundswell Approach and provides the scientific community and media coverage valuable data and images that may have been overlooked throughout the world.

In the last chapter on the Groundswell Approach and the first half of this book, we examine a series of case studies from around the world when youth were given "a seat at the table". Beginning with a 2017 study conducted by Canada's Organisation for Economic Co-operation and Development (OECD), we see the mutual benefits of participatory policy creation for both government officials and youth climate activists. While well-intentioned, the report warns of agendas set by the policymakers. Youth agendas set by adults who hold this opinion might focus on young people's shortcomings and problems rather than invest in youth's potential as a positive source of change (OECD 2017). How this can be overcome is explored through an examination of a variety of strategies presented by Luke Spajic, Georgia Behrens, Sylvia Gralak, Genevieve Moseley, and Daniel Lindolm in their 2019 commentary piece for *The Lancet* entitled "Beyond Tokenism: Meaningful Youth Engagement in Planetary Health".

One of the strategies for meaningful collaboration between youth and policymaker comes from South Africa in its Youth Climate Action Plan. The national framework is built upon "five core pillars": *Intersectionality; Advocacy and Activism for Climate Action; Good Governance; Systemic Change; Environmental Sustainability.*

Another strategy from the Netherlands proposes the creation of a series of position papers and manifestos created together with youth: "By entering into discussions on behalf of our supporters with politicians, policymakers and the business community, we give young people a voice in their own future", according to the Dutch youth climate organization *Jonge*

[1] https://pulsespikes.org/story/zero-hour

*Klimaatbeweging* (Youth Climate Movement). These documents are then shared with partner countries Uganda, Bangladesh, Chile, Egypt, Mexico, Nigeria, South Africa, India, and Qatar so their governments and youth climate activists can use them as models for their own policies in yet another example of the Groundswell Approach to establishing progressive climate policy with youth having a seat at the table.

The conclusion of the first part of this book details a proposed framework for using the Groundswell Approach by youth climate activists. The PLAN model stands for **P**rotest Protocol, **L**eadership, **A**ction, and **N**exus. This recommended infrastructure takes into account the successful elements of the Groundswell Approach and offers strategies that could address the shortcomings currently experienced.

Chapter 5 represents the introduction of a proposed framework informed by the successes and challenges of the Groundswell Approach. It takes into account the methods and strategies of engagement with government officials in a more straightforward way: the *Direct Approach*. This model proposes five components that need to be present for representative, inclusive, and ultimately successful climate policy creation in full participation with youth. The five areas are:

1. *Access*: Members of the global community of youth need to know that invitations and opportunities exist for direct engagement with policymakers. Access to these opportunities must be provided by the UN through channels familiar to youth for meaningful discourse.
2. *Respect*: Both parties need to enter policy collaboration with an open mind since neither party is fully aware of each other's perspectives and experiences. Efforts to learn and understand these defining characteristics will facilitate discussions. Regarding youth as a separate and distinct global community, as well as its inherent sub-cultures and differences within itself, acknowledges a culture that the policymaker needs to consider for fully inclusive and representative climate policy.
3. *Collaboration*: Youth must be prepared to conduct research and present reliably researched evidence that may be absent from the policymaker's understanding of the community's unique relationship with the impacts of climate change. The policymaker, in turn, needs to be open to assigning such knowledge collection and accepting it as reliable representations of that community's culture.

4. *Finance*: Funding programs and relative mechanisms need to be provided to supplement opportunities for direct engagement. Many young people cannot afford the travel to the policymaker's offices and many policymakers do not have a specific budget to travel to various and remote communities of youth leadership. It therefore becomes apparent that UN-sponsored funding programs need to be established to facilitate mutually agreeable places for engagement.
5. *Authorship*: Youth need to be able to contribute directly to policy writing and feel they have some acceptable degree of ownership with the drafted policy. Review and approval phases for youth would contribute to this. This component ensures that the voice of youth is not only heard, but is accurately represented in any ensuing policy.

Informing this framework are three previous models proposed at different times in the lifespan of contemporary youth activism: Roger A. Hart's "Ladder of Participation" (1992) prepared for and published by UNICEF; Harry Shier's "Pathways to Participation" (2001); and Helen Cahill and Babak Dadvand's "7 Ps of Participation" (2018). Shier's Pathways model was inspired by the United Nations Convention on the Rights of the Child's Article 12 which provides for state parties to acknowledge and make allowances for participation of youth in policy matters that impact on them. With these two UN-inspired and produced models, Cahill and Dadvand expanded these frameworks to include seven areas, all beginning with the letter "p", to advance the participatory opportunities and ideology of youth involvement with policy creation. The "7 Ps" are:

1. Purpose
2. Positioning
3. Perspectives
4. Power Relations
5. Protection
6. Place
7. Process

The 7 Ps is a valuable model of youth participation with political power as it incorporates the strengths of the previous models proposed by Hart and Shier, and fleshes out some of the considerations suggested, but not directly addressed by the previous models such as Place and Power Relations. Collectively, all three models serve to inform a new proposed framework made in this book. It should be noted that the author is also informed by his own direct experiences attending the annual United Nations climate summits, the COP conferences, from 2009 to 2022. In this capacity, direct field research and observations are made with respect to youth representation and the use of film by youth as an effective communications tool in participating in these conferences with the policymaker.

Analysis of this model and the practical applications of them are examined through the extensive scholarship provided by Harriet Thew and her valuable work on youth and their relationship with the United Nations Framework Convention on Climate Change (UNFCCC). In particular, the Direct Approach is applied to three of her published works to illustrate the effectiveness of the method:

- *Youth Participation and Agency in the United Nations Framework Convention on Climate Change* (2018);
- *"Youth is Not a Political Position": Exploring Justice Claims-making in the UN Climate Change Negotiations* (2020); and
- *"You Need a Month's Holiday Just to Get over It!" Exploring Young People's Lived Experiences of the UN Climate Change Negotiations* (2022).

The UNFCCC experiences of both Thew and myself, in concert with the models of youth participation proposed in previous years by Roger A. Hart (1992), Harry Shier (2001), and Helen Cahill and Babak Dadvand (2018), collectively contribute to the philosophy of the Direct Approach proposed in this chapter and throughout this book.

Chapter 6 extends the investigation of how the Direct Approach is used between youth and the UNFCCC comparing it against the Groundswell Approach in terms of effectiveness in giving youth a participatory voice at the COP conferences. Political scientist Jens Marquardt casts doubt on the efficacy of the Fridays for Future protest marches acknowledging their success in sparking debate on taking urgent action on climate change, but pointing out that "the movement's broader societal and political implications are yet to be seen" (Marquardt, 2).

In this chapter we also hear from the policymaker who seems to side with the Direct Approach as a means of engaging with youth productively. A recent study of climate protests among young people in Canada found that Members of Parliament (MPs) found a *Direct Approach* would be "more effective". In their study *Can Citizen Pressure Influence Politicians Communication about Climate Change? Results from a Field Experiment*, Seth Wynes, John Kotcher, and Simon D. Donner argue that youth are not inclined to speak directly with power. Their survey of protestors attending the Fridays for Future rallies in 13 cities found that of the largest age group, teenagers, only 10% had ever contacted a government official. The study recommends more direct contact with government officials since "these interventions are expected to be more effective" (Wynes et al., 17).

The United Nations echoes these sentiments with their call for more direct engagement with UN policymakers in a 2013 report entitled *Climate Change: Children's Challenge*. The report lays out a series of recommendations for youth involvement in policy creation and topping this list is the recommendation that "(Youth) should be involved in the development of the climate change policies that will affect their lives through inclusive consultation mechanisms" (Burgess, 3). The report goes on to state a claim many youth make today, nearly ten years after the report was written, and that is that youth "will inherit the outcomes of today's climate change decisions. It has no voice in the process. Adults, who will not necessarily have to live with the implications of their decisions, are making climate change policies without consulting or listening to (youth)" (Burgess, 9).

Local, national, and international governments all seem to agree that there are benefits to involving youth in climate change policy processes, but few seem to have a mechanism in place to actually make this happen. An investigation into youth participation at COP26 in Glasgow, Scotland, in 2021 found that many youth felt "excluded" from the process and "betrayed" by UN youth groups such as COY (the Conference of Youth) for arranging activities outside of the COP convention centre rather than inside. One of the fundamental areas of criticism by youth in Glasgow is that the four major components of the Conference of Youth fall short of actual engagement with UN officials.

The four components are:

1. (Creating a) Policy Document;
2. Capacity Building;
3. Skill-building Workshops; and
4. Cultural Exchange (Conference of Youth 2021).

Critics of COY claim the components do not engage the delegates of COY directly with the policymakers of COP. Rather, they tend to suggest COY participants need to learn (items 3 and 4) and network (item 2) and practice writing their own policies (item 1). None of these components addresses collaboration or meaningful policy representation with world leaders, their policymakers, or other United Nations officials who work on UNFCCC policy and communications.

The chapter explores governments who are trying to adhere to the framework of the Direct Approach by establishing departments and agencies specifically designed to engage youth in climate policy talks. Canada's Youth Climate Lab, Canada's Student Energy, and its own Youth Policy find ways of engaging youth not only with government officials, but with leaders of business and industry to voice their concerns on practices that exacerbate climate change. Other countries such as England and Ireland are also examined for their youth initiatives as a preface to the three main youth groups of the UNFCCC: YOUNGO, the Conference of Youth (COY), and Action for Climate Empowerment (ACE). Since the conference has put restrictions on non-state actors participating directly in policy negotiations in the Blue Zone of COP, these UN agencies are developing ways of involving youth nonetheless. Programs such as the *Global Youth Video Competition*, *Earthbeat*, and the *Youth Climate Report* are all examples of the UNFCCC's attempt to involve youth through a citizen science and citizen journalism approach using the medium of film that the UNFCCC has come to adopt as a reliable data delivery system and one that today's generation of youth have great facility in using.

Based on the programs and preferred methods of engagement identified in Chap. 6, the final chapter of this book looks at how and why film has become such an effective took for both youth and UN officials alike. After introducing film as a data delivery system in 2009 at the COP15 climate summit in Copenhagen, with the film *The Antarctica Challenge: A Global Warning* (2009), delegates attending the conferences found that the medium of film and its genre of documentary helped contextualize scientific reports for UN policymakers. It also provided valuable visible evidence of climate impacts. The following year, another feature documentary film, *The Polar Explorer* (2010), was screened at the COP16 conference in Cancun, Mexico. Here the film contributed directly to a resolution included in the Cancun Accord: *Enhanced Action on Adaptation: Section II, Subsection 25* (UNFCCC 2010). The success of film in policy negotiations and the recent call for expanded youth participation prompted a call from the UNFCCC for the creation of a project that allowed the global community of youth to submit filmed "reports" to the COP conferences each year. In 2011, the *Youth Climate Report* (YCR) was formed by myself and Toronto environmental advocate John Kelly. Beginning as an anthology film of five to six videos stitched together as a one-hour feature that was presented each year up to and including the Paris climate summit in 2015, the project responded to the demand for more content by remediating the documentary feature film format to a database documentary film format presented on a platform of a geographic information system map of the world. After a few enhancements requested by the UNFCCC, the new project was officially adopted in 2016 under the UNFCCC's Article 6 mandate as a data delivery system for the COP conferences.

This chapter examines this phenomenon and YCR's relation to youth through analyses provided by such eco-scholars as Helen Hughes, Scott MacKenzie, and Anna Westerstahl Stenport. The chapter also examines the growing phenomenon of "feeder programs", several workshops, training sessions, and competitions to assist young people in making these short three- to five-minute long video reports. Films made through these programs are added to the YCR project each year.

This book endeavours to identify how the United Nations is responding to the global community of youth's demand for urgent climate action. This book also looks to see how youth are evolving in their approaches to engage with these international environmental policymakers. There are clear advances being made on both sides and the path to participatory politics becomes shorter for the next generation of global citizens looking to end climate change before it ends them.

## Works Cited

"Article 2.1 (a)", *Paris Agreement*, United Nations, 2015. Accessed: May 29, 2022. Link: https://unfccc.int/sites/default/files/english_paris_agreement.pdf.

*Cancun Adaptation Framework.* Cancun: United Nations Framework Convention on Climate Change, March 15, 2011. Web. Accessed May 5, 2022. Link: https://unfccc.int/files/adaptation/cancun_adaptation_framework/adaptation_committee/application/pdf/1_cp.16.pdf.

Arocena, Rodrigo & Sutz, Judith. "Latin American Universities: From an Original Revolution to an Uncertain Transition", *Higher Education* (2005), Universidad de la Repu´blica, Uruguay, pp. 573–592.

Coburn, Oliver. *Youth Hostel Story.* National Council of Social Service, 1950.

Elmer-Dewitt, Philip. "The Ecokid Corps", *TIME Magazine.* New York: December 24, 1990.

Kern, Edward. "Can It Happen Here?", *LIFE Magazine*, October 17, 1969.

Macleod, David I. *Building Character in the American Boy: The Boy Scouts, YMCA, and Their Forerunners, 1870–1920.* University of Wisconsin Press, 2004.

Mandelbaum, Michael. "Vietnam: The Televised War", *Daedalus*, Vol. 111, No. 4, Print Culture and Video Culture (Fall, 1982), pp. 157–169.

Mazzini, G. 1926. 'Manifesto del comitato centrale democratico Europeo agli Italiani' [1851]', Scritti Editi ed Inediti di Giuseppe Mazzin, Vol. 46. Imola: Paolo Galeati.

Mazzini, Guiseppe. *Joseph Mazzini: His Life, Writings, and Political Principles* (New York: Hurd and Houghton, 1872), pp. 71–74.

Milanesio, Natalia. "Gender and Generation: The University Reform Movement in Argentina, 1918", *Journal of Social History*, Winter, 2005, Vol. 39, No. 2, *Kith and Kin: Interpersonal Relationships and Cultural Practices* (Winter, 2005), Oxford University Press, pp. 505–529.

Murphy, Patrick D. "Speaking for the youth, speaking for the planet: Greta Thunberg and the representational politics of eco-celebrity." *Popular Communication* 19, no. 3 (2021): 193–206.

Murray, Lorraine, ed. *Italy.* New York: Brittanica Educational Publishing, 2014.

Nagy, Gabriella. "The Nexus between Hiking and Youth Hostels (1907–1933): A Historical Analysis of the Evolution of the German Youth Hostel Movement." *Journal of Outdoor Recreation and Tourism* 23 (2018): 59–66. doi:10.1016/J.JORT.2018.07.006.

Robertson, Priscilla. "Students on the Barricades: Germany and Austria, 1848", Political Science Quarterly, Vol. 84, No. 2 (Jun., 1969), pp. 367–379.

Savage, Jon (2008). *Teenage: The Prehistory of Youth Culture: 1875–1945.* London: Penguin Books, 576 pp.

Terry, Mark. *The Geo-Doc: Geomedia, Documentary Film, and Social Change.* Basingstoke, UK: Palgrave Macmillan, 2020.

UNFCCC. (2010). *The Cancun Agreements*, Decision 1/CP.16, Report of the Conference of the Parties on its sixteenth session, held in Cancun from 29 November to 10 December 2010 (FCCC/CP/2010/7/Add.1). Bonn: United Nations Framework Convention on Climate Change Secretariat.

Williams, John Alexander. "Ecstasies of the Young: Sexuality, the Youth Movement, and Moral Panic in Germany on the Eve of the First World War", *Central European History* , 2001, Vol. 34, No. 2 (2001).

Zwicker, L. F. (2009). The Burschenschaft and German Political Culture, 1890–1914. Central European History, 42(3), 389–428. https://doi.org/10.1017/S0008938909990033.

PART I

# The Groundswell Approach

Youth climate activists from the Youth Climate Report at a Fridays for Freedom school strike march in Toronto, Canada. (Photo Credit: Youth Climate Report © 2018)

CHAPTER 1

# Disruptive Voices

> My generation has largely failed until now to preserve both justice in the world and to preserve the planet. It is your generation that must make us be accountable to make sure that we don't betray the future of humankind.—*United Nations Secretary-General, António Guterres*

A youth movement can be defined as an intentional attempt by young people to disrupt the status quo within a society or community for the purpose of eliminating a perceived problem and, by doing so, improve that society or community for the benefit of all its members. Historically, the methods by which young people have attempted these disruptions have not always been organized. Recognizing these shortcomings, these attempts to change society have also evolved over the years with new approaches, enhanced media coverage, and extended reach due to substantial advances in telecommunications technology and access to it. These become important transitions to consider when the youth of today are uniting over global issues like climate change. Globalization is not only becoming more easy through technology, but more encouraged by institutions, businesses, and with general social constructs. This establishes fertile ground for the youth of today to realize the goals of their predecessors and effect social change on a global scale.

There are two ways of achieving global climate policy change that youth are engaged in that this book will examine. These are what I refer to as the


M. Terry, *Speaking Youth to Power*, Palgrave Studies in Media and Environmental Communication,
https://doi.org/10.1007/978-3-031-14298-7_1

*Groundswell Approach* and the *Direct Approach*. The former has a rich history in mass gatherings, protest marches, flash mobs, all with varying degrees of success as youth experimented with these methods, often failing, as we will see, due to poor organizational skills. Their methods of recruitment, for example, have come a long way from flyer distribution in the 1960s to the viral social media event announcements of today and this section of the book concludes with a proposed framework for effective global strategies using this approach.

The Direct Approach proposes another new framework for youth engagement with power. Setting aside the mass gatherings, this method has shown success in terms of direct impact when those in power are seen as collaborators and not as enemies.

Before we examine the proposed frameworks, methodologies, organized structures, and impacts of youth movements in disrupting the norms that have paved the way to engaging politically in the climate change crisis, we first need to study how they got there and what twentieth-century youth movements empowered, emboldened, and encouraged young people worldwide to demand the global systemic change to save the planet.

As the first generation to actively campaign on a global scale against political inaction related to climate change, young people in the twenty-first century have a vested interest in effecting progressive and systemic change in environmental policy worldwide. The generations before them are largely responsible for these environmental shifts, but seem uninterested or unable to enact sweeping changes to global industries, behaviours, and legislation that would stem or reverse what scientists are predicting to be catastrophic calamity to our planet that will impact the lives of everyone.

For this introductory chapter, an attempt will be made to condense the past 170 years of youth movements throughout the world before focusing on specific youth activities in the 1960s in the United States as this period signals the birth of global youth movements on issues of international concern through the introduction of communication technologies. Specifically, television and the Internet helped establish what Marshall McLuhan once described as the "global village" (McLuhan 1962, 31). Together, the historical review and the 1960s' overview will provide foundational understanding of what motivates youth to protest in large, communal groups and how the Groundswell Approach yields the results they seek and when they don't.

## 1.1 GERMANY: THE BURSCHENSCHAFT MOVEMENT (1815 TO 1848)

In Germany, between 1815 and 1848, a national student association known as the *Burschenschaft* ("fraternity") *Movement*, flourished as a group of university students. It is often seen among scholars as both the first modern student movement and a clear example of modern generational conflict between old and young (Stark 1978, 323). The youth of this movement are characterized by their unconventional attire and their radical rejection of the values of their elders, and as such, are often seen as one of the first versions of the 1960s' hippies of the United States.

Following the end of the Napoleonic Wars in 1815 that saw German youth volunteer for duty, many of these young soldiers were not content to simply return to their studies at university. Five fraternities at Jena University dissolved on May 29, 1815 and reformed as one on June 12, 1815, with the name of Burschenschaft (Zwicker 2009, 399). Although claiming no political allegiance, the group of students claimed to be all-inclusive.

> …they were in favor of democratic integration of all social classes, and they hoped their practice would be a model for all German citizens. They proposed to further this aim through a strict code of personal morality. (Robertson 1969, 368)

One of the first gatherings of youth from this new student group was at an event they organized called the Wartburg Festival on October 18, 1817 at the Wartburg castle near Eisenach in Thuringia. Buoyed by a sense of national unity and pride following their victory over Napoleon, the students organized the festival to commemorate the 300th anniversary of Martin Luther's nailing of his 95 theses on the front door of the Castle Church in Wittenberg, on October 31, 1517. The event was marked by speeches in the castle's great hall and book burning.

> Four hundred students from a dozen institutions enjoyed a torchlight parade and a huge bonfire in which they burned symbols both of foreign domination and of domestic repression – a corporal's cane, a wig, a cop of *Code Napoleon*. They were forging a new brand of liberalism, different from French revolutionary and German feudal notions. (Robertson 1969, 368–9)

The event helped galvanize the youth of Germany and define how they saw themselves and how they saw their elders.

> Unlike traditional student culture, they demanded that their members focus on their studies and integrate intellectual interests with sociability and patriotism. Burschenschaft members particularly distinguished themselves from other fraternities by their willingness to engage in political education and with a political stance that was Protestant, liberal, and German nationalist. They stood against absolutism and against the remnants of feudalism that remained within the German states. Thus they were one of the first nationalist organizations that crossed German-speaking central Europe. (Zwicker 2009, 399)

It is important to note that these university students were content with the prevailing curriculum and educational system of German universities and were not intent on post-secondary reform as we will see was taking place at the same time in Argentina. Instead, these university students formed an allegiance to better define themselves in terms that represented how they saw themselves, not as junior versions of German adults, but as distinct members of the same society.

## 1.2 ITALY: Young Italy (1830s)

The next significant youth movement came in 1831 in Italy and was also nationalistic in nature with an aim to unify all Italians. Known simply as *Young Italy*, it was founded by revolutionary Giuseppe Mazzini, a doctor's son who showed great intellectual promise by entering the University of Genoa at the tender age of 14; 8 years later he graduated with a law degree. He is perhaps best known for his definition of democracy so often misattributed to American history: "a republic of the people, by the people, and for the people", written in 1851 (Mazzini, 104).

Unlike most youth movements in the late nineteenth and early twentieth centuries, membership in Young Italy required the taking of an oath. The oath was administered by an "initiator", an existing member who recruited new members. New members recited the oath in the presence of "the initiator: in the name of God and Italy" (Mazzini, 1831). This is the oath in its entirety:

> In the name of all the martyrs of the holy Italian cause who have fallen beneath foreign and domestic tyranny; By the duties which bind me to the land wherein God has placed me, and to the brothers whom God has given me; By the love—innate in all men—I bear to the country that gave my mother birth, and will be the home of my children; By the hatred—innate in all men—I bear to evil, injustice, usurpation, and arbitrary rule; By the blush that rises to my brow when I stand before the citizens of other lands, to know that I have no rights of citizenship, no country, and no national flag; By the aspiration that thrills my soul towards that liberty for which it was created, and is impotent to exert; towards the good it was created to strive after, and is impotent to achieve in the silence and insolation of slavery; By the memory of our former greatness, and the sense of our present denigration. By the tears of Italian mothers for their sons dead on the scaffold, in prison, or in exile; By the sufferings of the millions. I,...believing in the mission intrusted by God to Italy, and the duty of every Italian to strive to attempt its fulfillment; convinced that where God has ordained that a nation shall be, He has given the requisite power to create it; that the people are the depositaries of that power, and that in its right direction for the people, and by the people, lies the secret of victory; convinced that virtue consists in action and sacrifice, and strength in union and constancy of purpose: I give my name to Young Italy, an association of men holding there same faith, and swear: To dedicate myself wholly and forever to the endeavor with them to constitute Italy one and free, independent, republican nation; to promote by every means in my power—whether by written or spoken word, or by action—the education of my Italian brothers towards the aim of Young Italy; towards association, the sole means of its accomplishment, and to virtue, which alone can render the conquest lasting; to abstain from enrolling myself in any other association from this time forth; to obey all the instructions, in conformity with the spirit of Young Italy, given me by those who represent with me the union of my Italian brothers; and to keep secret of these instructions, even at the cost of my life; to assist my brothers of the association both by action and counsel—NOW AND FOREVER. This is I do swear, invoking upon my head the wrath of God, the abhorrence of man, and the infamy of the perjurer, if I ever betray the whole or a part of this my oath. (Mazzini, 1831)

There are some obvious themes here of family and religious faith that dovetail nicely with allegiance to the disruptive anarchistic political intentions of its founder, Giuseppe Mazzini. The oath was intended to solidify loyalty and commitment to the cause by invoking other unquestionable areas of fidelity in the lives of the young people at the time. Clearly political in nature, its charismatic leader was successful at recruitment growing the movement in numbers from 40 in 1831 to more than 50,000 just two years later (Murray 2014, 209); however, he was quite unsuccessful in mobilizing these apolitical youth.

> Young Italy plotted conspiracies against the existing governments in Italy during the 1830s and 1840s, but its revolts met with failure. The lack of popular support for insurrection as the road to independence discredited the society. (Murray 2014, 209)

After a series of failed coup attempts in Sicily, Abruzzi, Tuscany, Lombardy-Venetia, Romagna, and Bologna, it became clear that the members of Young Italy were either incompetent or uninterested in mobilizing their political views to such extreme lengths. In this case, youth gathered in large numbers not as a benevolent fraternity, but as an amateur army because of a leader with a passionate personality that seemed to have their best interests at heart. His devotion to the movement was a devotion to its members; however, as we will see in the 1960s in the United States, violence dissuades even the most idealistic youth.

## 1.3 GERMANY: WANDERVÖGEL (1901 TO 1913)

The turn of the twentieth century saw a different kind of youth movement, more focused on children than young adults. Wandervögel ("wandering bird") was a youth movement founded in Germany on November 4, 1901 by Karl Fischer. Established seven years before Robert S. S. (later Lord) Baden-Powell wrote the book *Scouting for Boys*,[1] Wandervögel aimed to give its members opportunities to explore life outside of urban environments and to commune with nature. While the Boy Scouts had a rigid age range of between 12 and 18, Wandervögel's membership was significantly younger. In fact, "only half of Wandervögel was (in this age range); better than eleven percent were nineteen or older" (Macleod, 282).

Unlike earlier European youth movements, this one was not founded by university students like the Burschenschaft Movement, nor was it established by a charismatic young anarchist with political reform intentions, but rather by a former student who liked the idea of school field trips. The Wandervögel Committee for Student Trips was first established by a group

[1] Baden-Powell, Robert. *Scouting for Boys.* London: H. Cox, 1908.

of former stenography students led by Fischer. Always seeking permission from the schools the boys attended, and with a strict set of rules to follow, the committee organized multi-day field trips.

The Wandervögel developed a symbolic language for their life on the road, they hiked through forests and over mountains, carrying all of their provisions on their backs, they sang folk and travel songs, cooked in the open air, and slept in villages, inns, hay barns, and tents (Coburn 1950). The "field trip" status soon became classified as a "youth movement" when many adults of the day disapproved of these expeditions and activities, especially when girls began to join in the expeditions.

> …a large part of the population considered the social order with its norms and traditions to be at risk from this revolutionary youth, thus creating a negative attitude towards them. Despite this, the movement was successful from the very beginning because it had the support of many young adult teachers and reformist school authorities, as well as scientists and writers, who recognised the importance of physical education and pedagogical innovation. (Nagy 2018, 61)

The movement soon evolved into segments, groups, and branches that went on trips without adult supervision. Generational independence soon became one of the defining features of the Wandervögel movement, now seen more as a *youth* movement.

> All members of the youth movement were committed to the Wandervögel principle of youthful freedom and sociability in peer groups relatively independent of adult control. In their privileged social position as sons and daughters of educated bourgeoisie, the Wandervögel were able to fashion a subculture based on the idea of generational independence. Most importantly, they developed a neoromantic conception of hiking into a natural landscape that they considered liberating and ennobling. Hiking was also seen as a practical method of self-education to physical toughness, self-discipline, and love of the homeland. The(ir) motto 'work on yourselves, make your hikes more and more exemplary' became a key tenet of youth movement ideology. (Williams 2001, 166)

Popularity in the movement increased throughout Germany and across its borders. New versions of Wandervögel were emerging with less structure and with different goals. One extreme incarnation gathered in the Swiss village of Ascona and "created a commune that practiced vegetarianism, nature cures, and free love" (Savage 2008). The original objective of hiking turned into escaping adults and their urban environments. The goal of communing with nature was interpreted by establishing a permanent residence in nature. This "constituency of the sensitive" offered a safe haven for anarchists, intellectuals, and Wandervögel alike (Savage 2008). It also gave us a first look at an anti-establishment, nature-based, hedonistic, apolitical youth that would be defined as "hippies" 60 years later.

## 1.4 ARGENTINA: The Argentine University Reform (1918)

This event marked a youth-led call to action that yielded specific impact and change on a youth-centric domain. It differs from the other examples of gatherings of youth in large numbers whose aims were a rebellious carefree lifestyle or a general government reformation. In this case study, we see how young people worked within the system instead of outside of it to establish university reform for themselves and for future generations of youth.

With three universities established by the Jesuit order of the Roman Catholic Church at the time, the curricula of these post-secondary institutions were restricted to the belief systems of this faith. Progressive ideas introduced by European immigrants often did not make it to these classrooms. The universities in Argentina founded, owned, and operated by the Jesuits were:

- University of Buenos Aires (Buenos Aires, 1821)
- Philosophical and Theological Faculty of San Miguel (San Miguel, 1918)
- National University of Córdoba (Córdoba, 1613)

With the first democratically elected president in office, Juan Hipólito del Sagrado Corazón de Jesús Yrigoyen (1916 to 1922 and again from 1928 to 1930), a member of the Radical Party, he ended 36 years of Conservative rule (Walter 1969, 234). Together with an influx of progressive ideas introduced by waves of European immigration at the time,[2] the climate seemed right for young people to demand university reform.

[2] The 1895 census counted four million Argentines; by 1914, this figure had doubled. Much of the increase represented the more than two million European immigrants who entered and remained in Argentina during this period (Walter 1969, 234).

The University Reform Movement was organized by students at the National University of Córdoba who, in March of 1918, rebelled against the Jesuit-run university "accusing the professors of being authoritarian, inefficient, clerically oriented, and obscurantist" (Milanesio 2005, 505). More specifically, the students rebelled against the anti-secular teachings and nepotism among professors that prevented a more liberal and inclusive body of study. Many students in neighbouring, Catholic-run universities embraced this ideology and soon the movement spread to universities in Buenos Aires (1918) and La Plata (1919–1920) putting pressure on the national government for sweeping legislative reforms (Milanesio 2005, 505). Collectively, this group of youth adopted strategies that ultimately proved successful.

> Through strikes, rallies, petitions to the national authorities and the seizure of the (National University of Córdoba) in September 1918, the reformists successfully forced the national government to carry out the University Reform. The distinctive nature of the movement derived not only from its radical demands, but also from its extremist tactics, the level of sophistication of its organization, and its major continental impact. In fact, the Reform Movement spread from Cordoba to Lima (1919), Cuzco (1920), Santiago de Chile (1920), and Mexico (1921)...on October 7, 1918, President Hipólito Yrigoyen approved a national decree that incorporated almost all of the reformist demands. (Milanesio 2005, 505–506)

But perhaps most notable about this student-led movement beyond its success in introducing new national educational legislation, was its far-reaching goal of democratising all of Argentina by introducing areas of study that would make the university itself a democratising agent. They achieved this by controlling the content of the university's curriculum to ensure "autonomous rule of universities by its students, teachers and graduates" (Arocena and Sutz 2005, 574) so external hegemonies like the government or the Church could never again exercise authoritarian control. In all chapters of the University Reform Movement, students established a framework to serve as a model for all universities based on three core principles: teaching, research, and extension. The teaching principle included areas of study not previously allowed and free tuition allowing access to all classes; the research principle included science that aided in national development; and the extension principle permitted the collaboration with the least favoured sectors of the population through cultural diffusion and technical assistance (Arocena and Sutz 2005, 575).

This serves as an excellent example of how youth mobilizing in a sector that is predominantly occupied by themselves yields success when employing the Groundswell Approach. In contrast to other youth movements that attempted to change government rule, end wars, start wars, or otherwise substantially change societal status quos, they were not met with as much success as they were when they applied their passionate forces towards areas of their own domain.

## 1.5 USA: Hippies and Yippies (1960s)

Not since the 1960s have we seen a youth movement so dedicated to establishing a counterculture necessary to overthrow what they see as a dangerously ineffectual and apathetically incompetent hegemony worldwide. But what we see around the world today in massive marches and protests represents a somewhat different approach by youth to disrupting the system. Seventy years ago, the "make-love-not-war" counterculture did not intend to dismantle the existing regime, but rather co-exist with it as an alternative culture of peace and hedonism. The youth at this time were called "hippies" (Cavan 1972) and their "drop-out" generation lifestyle is often traced back to San Francisco circa 1965 in the Haight-Ashbury district near Golden Gate Park (Davis 2015). These middle-class youth found community with each other living in this neighbourhood defined by an escapist approach to living outside the expectations of capitalism.

Unlike the youth climate movement of today, the hippies of the 1960s were neither connected globally, nor seeking to replace the dominant culture. Instead, they chose not to participate in their parents' world of structured commerce and social behaviour. What became a mantra of "sex, drugs, and rock 'n' roll" (Kern 1969, 77)[3] became the defining features of a generation that cared not to be defined. If the Western culture at the time was responsible behaviour exhibited by gainful employment, sexual repression, and sobriety, the hippies steadfastly embraced a counterculture of unemployment, sexual freedom, and intoxication primarily through the use (and some might argue the abuse) of marijuana and lysergic acid

[3] First coined in an October 17, 1969 *LIFE Magazine* article as "sex, drugs, and rock". It first appeared as "sex, drugs, and rock 'n' roll" in 1971 in a UK publication, *The Spectator*, and later popularized by British band Ian Drury and the Blockheads in 1977 in a song of the same name.

diethylamide (LSD). They were escaping the confines of the prevailing culture, not trying to exchange it with theirs.

Nevertheless, this counterculture was perceived as being disruptive to the norm and, therefore, a threat to it. If the youth of the day could co-exist within this bohemian lifestyle, a strong argument for the virtues of Communism—while at the same time illuminating the flaws of Capitalism—can be made. This was seen to disrupt the status quo and dismantle a structured way of life enjoyed by their parents' generation. The hippies were now being seen not so much as lazy, aimless, and harmlessly hedonistic, but structured and destructive, a perception strongly rebuked by a community that claimed no political agenda (Kern 1969, 80).

However, society's resistance to their carefree lifestyle gave rise to a self-awareness the hippies did not previously have. They now knew they had a voice, a certain amount of power they could yield that could result in significant change in society, changes that matched their ideology of peace and love. Initially apolitical, the hippies were suddenly a political force to be reckoned with. The cause they decided to embrace was the Vietnam War.

Who could argue that war, in non-specific terms, was not *bad*, especially a war for which there was no clear reason to fight? The Vietnam War (1959–1975) was often billed as the first "televised" war (Mandelbaum 1982) as video coverage was prevalent and ubiquitous on televisions in the USA and around the world. This also provided a global platform for youth never before seen. Not only was their anti-war message ("make love, not war") amplifying the voice of youth in a way that reached more people worldwide than any other youth-led protest in the past, but the nature of the medium provided a face to youth as well.

The hippies could no longer be ignored and the "establishment" at the time began to counter this counterculture with increased police task forces arresting hippies for illegal drug possession and vagrancy. There was some resistance, but the hedonistic philosophy of this group of young people left no room for any kind of organized structure that may have resulted in more successful efforts. Combined with the decline of the Vietnam War, the counterculture was losing its only political clout and the hippies soon began "selling out", joining the establishment for economic and personal stability.

One may look at the hippie movement ultimately as a failure, but it served future generations of youth as a model for protest, global communication, and disrupting the status quo. A new generation of youth looked at this failed attempt as a model for future protests and saw some foundational approaches they could embrace: strength in numbers and media coverage. This gave rise to what I call the *Groundswell Approach*. Usually beginning on college campuses, the youth of the 1970s and 1980s were now more organized and began adding a valuable component to their protests: publicity. At the grassroots level, they began plastering the walls of their colleges with posters promoting their cause and public gatherings. Leaders would introduce newcomers at these events with passionate speeches about their cause, often with megaphones. They also extended their reach locally by promoting their gatherings with flyers off-campus. And finally, they began to learn how to enlist the power of the media with more formal announcements in the form of press releases. The conventional journalistic media of newspapers, radio, and television became their new megaphone.

Still, these gatherings were acts more of preaching to the converted rather than effecting measurable change in the now growing capitalistic establishment. The hippies who "sold out" had bolstered the societal culture of gainful employment and the family unit. Now youth activists perhaps understandably felt they could fight city hall. Instead, their approach to protest was largely through the arts. Popular music—or as it was labelled at the time, "underground music"—spoke for the post-hippie youth, as did creative messaging through theatre and graffiti.

This new group of youth were transitionally called *Yippies* to identify their new approach, but also to pay tribute to the gains made by the predecessors. It was a term coined by Paul Karssner who co-founded the Youth International Party founded on December 31, 1967 along with Abbie Hoffman and Jerry Rubin in 1967 (Kassner, 156).

Much like the hippies before them, they were not taken seriously, often denounced for their "theatrical" approaches to attracting attention to their causes. Nevertheless, their performances proved popular by providing entertainment to the masses so their message could be delivered to many in a form more dynamic than mere public addresses. Creating a new form of theatre—often outside of the traditional theatre venue—their unique style was formed from a combination of performance theories incorporating New York's avant-garde art world, the concept of guerrilla theatre outlined by R. G. Davis of the San Francisco Mime Troupe, and the notion of Theatre of Cruelty created by Antonin Artaud.

Seen today as a precursor to the *flash mob*—a spontaneous gathering of people in a large public space like a mall or a train station for the purpose of musical performance—the Yippies' unique brand of public display usually presented the group's political agenda with a call to action. Three well-known examples were:

- the 1967 "March on the Pentagon" where future Yippie leaders performed an exorcism ritual at the Pentagon;
- the 1968 "Grand Central Station Yip-In" event that advertised for the Yippie movement;
- and the 1968 "Festival of Life" at the National Democratic Convention in Chicago where the Yippies nominated a pig as presidential candidate.

Each event was notable for its expansive *cast* of participants who were recruited before the days of social media by letter-writing campaigns, posters, and phone calls. These large groups often were joined by eager onlookers who were either sympathetic to the cause of the performance or simply wanted to be part of the show.

Referred to as *revolutionary action theatre*, an estimated crown of 35,000–50,000 Yippies dressed in costumes of gypsies gathered at the Pentagon's north parking lot in Washington. Armed with nothing more than drums, cymbals, and flowers, they begin their play's one-line chant "Out, demons! Out!" as they attempt to exorcise the demons of war from the government building (Mettler 2017).

In recruiting participants, Yippie co-founder and organizer of the event, Abbie Hoffman, declared in a speech, "We will dye the Potomac red, burn the cherry trees, panhandle embassies, attack with water pistols, marbles, bubble gum wrappers, bazookas, girls will run naked and piss on the Pentagon walls, sorcerers, swamis, witches, voodoo, warlocks, medicine men and speed freaks will hurl their magic at the faded brown walls…We shall raise the flag of nothingness over the Pentagon and a mighty cheer of liberation will echo through the land." (Mettler 2017). Nearly 700 had been arrested for various acts of civil disobedience at the event.

The following year saw another form of protest performance, this time indoors at New York's Grand Central Station. A handbill promoting the event described it as "a spring mating service celebrating the equinox…a back-scratching party, a roller-skating rink, a theater, with you, performer and audience". Thousands of these flyers were distributed at a press conference held at New York's Americana Hotel on March 17, 1968. The goal was to attract media more than participants as the event was set to take place just a few days later on March 22.

Once again, we see youth at the time eager to be heard, but lacking the planning skills to pull off an organized event. But this is precisely the approach proposed by Artaud's Theatre of Cruelty, specifically to "reject form and incite chaos" (Jamieson, P. 22). The aim of this brand of theatre was to involve the audience in the performance directly so as to force engagement actively rather than passively.

> Artaud sought to remove aesthetic distance, bringing the audience into direct contact with the dangers of life. By turning theatre into a place where the spectator is exposed rather than protected. (Jamieson, P. 23)

The Yippies achieved this goal. An estimated 6000 people gathered for the event and while they gathered in Grand Central Station, it was clear a performance was not rehearsed or even scripted. Climbing to the top of the information booth several youth began yelling such apparently random things as "Yippie!", "Long Hot Summer!", and "Burn, Baby, Burn!" (*Village Voice* 1968). The gathering audience members did, indeed, participate often chanting these lines in unison. After an hour of this, swarms of police descended on the performers and audience alike brutally attacking them with nightsticks.

> "It was the most extraordinary display of unprovoked police brutality I've seen outside of Mississippi," Alan Levine, staff counsel for the New York Civil Liberties Union, said at a press conference on Saturday. "The police reacted enthusiastically to the prospect of being unleashed." Levine reported seeing several people forced to run a gauntlet of club-wielding cops while trying to flee from what has been characterized as a "police riot" (*Village Voice* 1968).

While only 57 people were arrested (*Village Voice* 1968), significantly less than those arrested in Washington for the military exorcism performance, there was significantly more bloodshed as the New York police seemed both unprepared to deal with a mob and eager to physically attack participants with nightsticks. One reporter from the *Village Voice* covering the event, Don McNeill, described the melee as "a fire in a theatre".

> My credentials were checked twice, and I was allowed to pass. At that point, I was stopped a third time by two uniformed cops. They looked at my credentials, cursed the Voice, grabbed my arms behind my back, and, joined by two others, rushed me back toward the street, deliberately ramming my

> head into the closed glass doors, which cracked with the impact. They dropped me in the street and disappeared. My face, and my press card, were covered with blood. I went to the hospital to get five stitches in my forehead. (*Village Voice* 1968)

While Artaud's unique brand of participatory theatre was self-identified as *cruel* and intended to promote *chaos*, this level of violence was not anticipated by this poorly organized troupe of youth activists. It did, however, reveal that their message, as vague and unstructured as it may have been, was being heard by the masses and was seen as a threat by those in power.

Later that year, the third significant experiment with youth activist gatherings in the name of change came at the National Democratic Convention in Chicago held between August 26 and 29, 1968. The Yippies staged events they called the "Festival of Life" and the "Unbirthday Party" of President Lyndon B. Johnson (Blake 1992). Both events attempted to be peaceful with performances from folk singing groups such as Peter, Paul, and Mary, but the lessons learned in New York did not help in Chicago. Thousands of participants—labelled protestors at the time—were not prepared for the excessive presence and violence of both Chicago Police and the National Guard. The events were notably more organized with guest speakers scheduled and a greater presence of television news media, but the 10,000 attendees were far outnumbered by 24,000 police (12,000), US army soldiers (6,000), and members of the National Guard (6,000) (*The Guardian*, 1968).

Perhaps most significant in this evolution of youth activism was the presence of CBS, a US national television network broadcasting events live for its nightly news program. Dan Rather did his best to report live from the scene despite being assaulted several times by the police. News anchor Walter Cronkite responded to Rather's report with "I think we've got a bunch of thugs here, Dan". The national and subsequent international media coverage on television of the horrific events of that day were categorized by a government report by Daniel Walker as a *police riot*:

> [O]n the part of the police there was enough wild club swinging, enough cries of hatred, enough gratuitous beating to make the conclusion inescapable that individual policemen, and lots of them, committed violent acts far in excess of the requisite force for crowd dispersal or arrest. To read dispas-

> sionately the hundreds of statements describing at firsthand the events of Sunday and Monday nights is to become convinced of the presence of what can only be called a police riot. (Walker, 5)

The events of August 26–29, 1968 also gave rise to a slogan that has endured to this day: "The whole world is watching" (Taylor and Morris 2018), referring to the television coverage that broadcast the violence live to millions in the USA and around the world. This becomes a pivotal point in youth activism. The sporadic gatherings of the hippie and yippie movements were coordinated through relatively primitive means: posters, flyers, handbills, usually homemade with poor handwriting and photocopying. They were handed out on the street and glued to telephone poles. A far cry from the global reach of today's social media, not to mention the professional design and layout afforded by today's digital software.

The television became the first electronic communications tool with a global reach to provide visuals of youth gatherings and through the impartial lens of the fourth estate, still respected as such in the days before social media. Young people around the world began to see images of their peers at these violent gatherings on their local news programs and were inspired to stage similar marches and protests, more to protest "police riots" than anything specifically American like the war in Vietnam. One of the most high-profile Yippies of the era, Jerry Rubin, once famously said "media does not report news, it creates it. An event happens when it goes on TV and becomes a myth".

Rubin and the Yippies believed they could control the medium by staging elaborate and theatrical events that attracted the attention of the viewer with dynamic visuals, like a TV commercial. Stylistically, the excessively visual components of a TV commercial are the antithesis of the dry rhetoric of a news program. The Yippies believed if they followed the model of the TV commercial, their television appearances on news programs and talk shows would have a greater impact on audiences and compel a new generation of apolitical youth to join in the fun.

> Do you think any one of the millions of people watching the show (*Meet the Press*) switched from being liberal to a conservative or vice versa? I doubt it. One thing is certain though…a lot of people are going to buy that fuckin' soap or whatever else they were pushing in the commercial. (Hoffman 1968)

This approach yielded success in mobilizing large groups of youth to join in the protests of the Yippies whose political messages were primarily anti-war, anti-establishment, and anti-capitalism. However, the heavy resistance in the form of violent reactions from police and other law enforcement officials did not result in a nationwide growth of the movement as hoped. Indeed, the arrest and conviction of Yippie leaders Abbie Hoffman and Jerry Rubin in 1970 for intent to incite a riot while crossing state lines signalled an end to the movement.

Obviously, the 1960s does not represent the genesis of youth climate movements we know today, but one aspect of it does. A technological transition was introduced that caused the Groundswell Approach of youth movements to grow in numbers and to reach participants worldwide: advances in telecommunications. With the advent of television, the whole world was indeed watching, and youth began to use press conferences and mass media to bolster their numbers and amplify their voices. Social theorist George N. Katsiaficas remarks on this turning point and identifies 1968 as the year youth movements took on a global composition.

> In 1968…television, radio, and travelling spokespersons spread (student) movement(s) around the world as never before, synchronizing its actions and making the political generation of 1968 a truly international one. (Katsiaficas 1987, 41)

In this formative year, the multimedia-supported outreach of youth movements saw 60 countries reporting youth-led protests and disruptions: 15 countries in Europe, 15 countries in Latin America, 16 countries in Africa and the Middle East, 11 countries in Asia, and 3 in North America (Braungart and Braungart 1990, 158). However, few of these protests were related to issues concerning the environment.

This chapter has not covered all the various youth movements in the past 170 years. The ones highlighted here, however, represent the early indications of how youth gathered and the causes they felt were important. The Burschenschaft and Wandervögel Movements in Germany revealed a desire for a simpler and more nature-based lifestyle free of authority and supervision. Their rebellions were against the structure of their parents' urban environments and behavioural expectations. Both movements embraced spending time away from cities and adults to "find themselves" instead of becoming blindly conforming to versions of their parents and other adults. A precursor to the "drop-out" lifestyle of the American youth of the 1960s, the German youth of the nineteenth and early twentieth centuries did not inspire the hippies as much as they demonstrated a natural inclination for youth to define their own identity through nature by means that rejected the structure of the societies into which they were born and raised.

In the early 1900s we see youth movements in both Italy and Argentina that shunned hegemonic authority as represented by church and state or as the hippies classified them "the man" or "the establishment". It is evident through these movements that youth do not want to be told what to do by power structures within which they cannot participate, but instead of replacing these institutions with their own, they default to "dropping out" as a means of both eluding these power structures and establishing their own.

The Argentine University Reform movement shows that a more organized group of youth focused with a specific agenda of change can actually achieve significant and progressive impact on society through the demand for and the implementation of new national legislation. It also reveals that when successful, other groups of similarly minded youth in other countries will replicate the cause and its methodologies to introduce similar change in their respective communities.

Politically, all these movements seem to show a proclivity towards socialism and Communism reflecting the desire of youth for inclusivity and inter-dependence with each other and independence from authority. In 2021, the *Washington Examiner* reported that 28% of American youth have a "favourable view" of Communism and another 30% feel the same about Marxism (*Washington Examiner*, July 20, 2021). Washington's *National Public Radio* reports that a similar figure of 26% of youth aged 18–39 have a "favourable" view of socialism (*Sanders* 2015) and a study out of the Pew Research Centre reports that 49% American youth between the ages of 18 and 29 have a "positive" view of socialism. In 2019, a Gallup poll found that 49% of American Millennials viewed socialism positively versus 51% who still favoured Capitalism. Only nine years before that, the gap was much wider at 66% favouring Capitalism and only 48% favouring socialism (Saad 2019).

The political nature of youth is compromised to some degree by their exclusion in the voting process. Conversely, politicians have traditionally paid little attention to youth for the same reason. If they are too young to vote for me why bother listening to them? But recent research suggests that the youth of the 2020s is actually more engaged than ever before, but are simply feeling they cannot trust those in power. A teaching resource published by Public Broadcasting System (PBS) Education in the USA claims that poor voter turnout among youth who are of age to vote is not a sign of apathy.

Power is constantly being withheld from young people, which limits and binds the type of organizing and political involvement they have. Interpreting low voter turnout by young people as evidence of apathy ignores the structural and organizational obstacles to electoral participation many young people face. It also ignores the distrust many young people feel towards traditional institutions of governance. Feeling disconnected from a process that is viewed as ineffective is not apathy. Especially when one considers how young people have been failed by political parties, including those that claim to represent (them).

## 1.6 Environmental Victories (1970s, 1980s)

The environment, in general terms, has always attracted the attention of youth, as we have seen with Germany's Burschenschaft and Wandervögel Movements and with the American Hippie Movement, but it was not until the 1970s and the 1980s that we began to see youth protesting and marching on behalf of the environment on a united and global scale.

During this era, three significant environmental issues—global in nature and anthropogenically caused—were targeted by both science and the public: a pesticide known as DDT (dichloro-diphenyl-trichloroethane); toxic fumes caused by lead in gasoline; and ozone-depleting chlorofluorocarbons and hydrofluorocarbons (CFCs and HFCs). Careless industry practices were behind these crises which were seeing deaths directly caused by DDT and lead, not to mention the creation of the ozone holes over the polar regions caused by CFCs and HFCs threatening the lives of those who lived closest to Antarctica in Australia, New Zealand, and Argentina. But perhaps most surprisingly, campaigns around the world were successful in convincing governments to ban these practices and legally compel corporations to change their business practices. Young people began to see that the organized efforts of average citizens banding together in large numbers and armed with scientific research, that is facts, could effectively defeat previously perceived corporate giants and cause progressive change for the planet and its people.

These citizen protests established a framework that proved indefensible to the corporate powers that were causing these environmental problems: the use of scientific evidence as irrefutable proof of the dangers of these corporate practices and their allegiance with lawmakers. This objective data may be dismissed and discredited by industry, but not by governments and legislators who found no way around ignoring the evidence presented by their citizens.

### *1.6.1 DDTs*

Challenging industry and demanding changes to business practices that almost always adversely impacted on their bottom line was akin to fighting City Hall. It was an uphill battle at best, and frequently a loss at worst. But hope was presented in a significant way in 1962 when Rachel Carson wrote a book called *Silent Spring*. Dubbed "The Gentle Subversive" and an "improbable revolutionary" (Lytle 2007), Carson exposed industry's use of poisonous pesticides in her book and it is often seen as the beginning of the environmental movement. Being neither a scientist nor a journalist, her writing style was both accessible to the general public, and dismissible by the chemical industry she was criticizing.

> "First serialized in *The New Yorker* in June 1962, (*Silent Spring*) alarmed readers across America and, not surprisingly, brought a howl of indignation from the chemical industry. 'If man were to faithfully follow the teachings of Miss Carson,' complained an executive of the American Cyanamid Company, 'we would return to the Dark Ages, and the insects and diseases and vermin would once again inherit the earth'".

The book took Carson four years to write and part of that time was spent building a case against the anticipated backlash from the chemical producers of DDT, the poisonous chemical found in the pesticides of the time.

> Anticipating the reaction of the chemical industry, she had compiled *Silent Spring* as one would a lawyer's brief, with no fewer than 55 pages of notes and a list of experts who had read and approved the manuscript. Many eminent scientists rose to her defense, and when President John F. Kennedy ordered the President's Science Advisory Committee to examine the issues the book raised, its report thoroughly vindicated both *Silent Spring* and its author. (NRDC 2015)

The subsequent banning of DDT raised an awareness that human intervention, especially on an industrial scale, could cause adverse effects to the environment and the people who live in it. According to the Natural Resources Defence Council (NRDC), this moment began the environmental movement: "For the first time, the need to regulate industry in order to protect the environment became widely accepted, and environmentalism was born" (NRDC 2015). Shortly after this ban, the United States establishes the Environmental Protection Agency on December 2, 1970 (*EPA History*). It can also be said that this marked an awareness of man's anthropogenic influence on the global ecology of the planet.

But perhaps most significant to future youth climate movements was the example Carson set as being able to do what was previously thought to be impossible: take on Capitalism and industry (the "Establishment" and the "Man", respectively) and win. This encouraged a new generation of youth to interrogate and challenge business practices when the scientific community raised environmental concerns.

### *1.6.2 Leaded Gasoline*

One of the first industry-versus-environment practices dates back to 1921 when lead (specifically, tetra-ethyl) was tested as an additive to gasoline (or petrol) to counter consumer complaints of engine "knocking" in their cars. This was done by a lab technician named Thomas Midgley at General Motors in Dayton, Ohio in the United States (Kovarik 2011), despite lead being known as a poison at the time. After his death, Midgley was described as being "responsible for more damage to Earth's atmosphere than any other single organism that has ever lived" (Walker 2007).

The poison proved to be successful in eliminating the issue of engine "knocking" and was commercially released. In 1923, the US Public Health Service (PHS) was made aware of this and began to ask for safety tests (Kovarik 2011). Despite several tests that proved lead was a contributing factor in the illness and death of those exposed to it, in 1926 the PHS finds "no good grounds" for banning ethyl gasoline, yet still called for further testing. Apparently, this edict fell on deaf ears as no independent studies were conducted until the 1960s (Kovarik 2011). In 1965, the first peer-reviewed research showed that "high lead levels in industrial nations are man-made and endemic" (Patterson 1965).

In 1970, the newly formed Environmental Protection Agency ordered car manufacturers to begin building engines to run on unleaded gasoline. The EPA ordered the introduction of unleaded gasoline and issued an order for lead-free gasoline cars by 1975, thus beginning the world's first national phase-out of leaded gas (Murdock 2005).

### 1.6.3 *The Ozone Hole*

On the other side of the world, a hole in the atmosphere over Antarctica caused by vanishing ozone gas was discovered in 1985 by Jonathan Shanklin, a junior researcher at the British Antarctic Survey. It was believed at the time that the Concorde supersonic jet was causing this depletion in earth's ozone layer so he set out to disprove this theory.

> There had been concern at the time that exhaust gases from Concorde (the supersonic passenger aircraft), or chlorofluorocarbons (CFCs) from spray cans, might damage the ozone layer. Being an ignorant physicist, I thought this unlikely, so decided to present that year's data and compare it with values my boss had computed from a decade earlier. I expected them to be the same, so Concorde would be able to keep flying and the public could keep using their spray cans. The only problem was, they weren't the same. ("The Story Behind the Discovery of the Ozone Hole" 2021)

Once the cause of this environmental crisis was identified the world was quick to act. Just two years later in 1987, the United Nations gathered in Montreal and signed an agreement to ban the industrial use of CFCs and hydrofluorocarbons (HFCs), most commonly found in spray cans. The Montreal Protocol was eventually signed by every single country in the world, the first time every nation on earth signed an agreement. What made this unilateral agreement additionally unique was how easily and quickly business and industry sided with the scientists and the governments of the world to ban a substance that they relied upon to a large degree. The Montreal Protocol is a shining light to today's young climate activists as it serves as an example that no matter how insurmountable an environmental crisis may seem to be, the global communities of government, business, and science can find a solution and agree to it within urgent timeframes.

Global or even national youth protests were not prevalent in the environmental issues of DDT, leaded gasoline, and the creation of the ozone hole, but these incidents marked the beginning of environmental activism among youth who now could see they needed to pay closer attention to what the industries and governments of older generations were doing to the planet they were soon to inherit.

The EPA ruling on leaded gasoline, together with the government ban on DDT and CFCs as well as the creation of the Environmental Protection

Agency, fostered a new suspicion of industry among Americans and, by extension, the world. With the depletion of the ozone layer caused by CFCs and HFCs, greenhouse gases emitted by industry worldwide, the public was now more untrustworthy of corporate giants in particular and Capitalism, in general, than ever before. This laid the groundwork for young people to question this branch of societal authority with confidence especially regarding corporate decisions that may directly impact on the environment and their own lives. As well, seeing how a book by a "regular citizen", Rachel Carson, and the subsequent bans on DDT, lead, and CFCs, youth were emboldened like never before to question and challenge industry. As telecommunications technologies advanced, youth began to follow the inspiration provided by the Yippies to protest in large numbers, but having learned the lessons of the importance of organization and the threat of police violence, tailored their new-found environmental activist intentions in more productive and safer ways.

## 1.7 Conclusion

We can construct a profile of youth activism based on the examinations made in this chapter. First, when so inclined, youth gathering together is frequently for the purpose of communal escape from a perceived authoritarian regime either at school, in their families, or in their cities driving them out in collective numbers to support each other in a simpler, rule-free environment commonly found in nature. Secondly, we can see a proclivity towards political change that often represents inclusivity, socialism, diversity, and other traditional liberal platforms to disrupt the power structure to which they cannot belong; all bets are off, however, when this resistance and rebellion is met with authoritarian violence. Thirdly, when a youth activist campaign is successful with measurable impact, other youth groups are eager to embrace the ideologies and methodologies employed to effect similar progressive change in their own communities. And lastly, electronic communication, both through traditional mass media (television, radio, print) and postmodern social media (websites, email, and networked platforms), not only extend the reach of youth and magnify their message, but provide the opportunity—with relative ease—of tackling global issues like climate change in unprecedented ways. A global response to a global crisis.

In the next chapter, we will examine the relatively new mobilization of this globalized community of youth as they gather in person and online in the millions to demand systemic changes to capitalistic structures that they believe are the cause of global warming and the subsequent changes to the climate that are threatening lives worldwide.

## Works Cited

"5 Held in Acid Rain Protests", *New York Times*, February 13. 1982. Accessed March 27, 2022. Link: https://www.nytimes.com/1982/02/13/us/5-held-in-acid-rain-protests.html.

*1968. A Report Submitted to the National Commission on the Causes and Prevention of Violence*. Bantam Books, 1968.

"Acid Rain Protestors are Active in Europe", *New York Times*, April 4. 1984. Accessed March 27, 2022. Link: https://www.nytimes.com/1984/04/04/world/around-the-wrld-acid-rain-protesters-are-active-in-europe.html.

"Dan Rather: A Reporter Remembers". *CBS News*. Archived from the original on March 2, 2012. Retrieved September 4, 2021.

"Engaging Young People in Politics", *PBS Teachers Lounge*. PBS Education, October 14, 2020. Link: https://www.pbs.org/education/blog/engaging-young-people-in-politics.

"EPA History", the United States Environmental Protection Agency, *epa.gov*. Accessed: March 29, 2022. Link: https://www.epa.gov/history#:~:text=Celebrating%2050%20years%20of%20the,1970%2C%20by%20President%20Richard%20Nixon.

*Little Change in Public's Response to 'Capitalism,' 'Socialism'*, Washington: Pew Research Center, December 28, 2011.

"The Story Behind the Discovery of the Ozone Hole", *UK Research and Innovation*, 2021. Accessed: May 6, 2022. Link: https://www.ukri.org/news-and-events/responding-to-climate-change/topical-stories/the-story-behind-the-discovery-of-the-ozone-hole.

*The Story of Silent Spring*. Natural Resources Defense Council (NRDC). August 13, 2015. Accessed: March 29, 2022. Link: https://www.nrdc.org/stories/story-silent-spring.

*Timeline*, August 28, 2017. Accessed: April 12, 2022. Link: https://timeline.com/yippies-grand-central-riot-e589f4f739df.

*Village Voice*, March 28, 1968. Accessed: April 17, 2022. Link: https://www.villagevoice.com/2020/10/06/the-grand-central-riot-yippies-meet-the-man.

Abodo, Sam. "Why do so many young people like communism?", *The Washington Examiner*, July 20, 2021.

Albert, Judith Clavir; Stewart, Edward Albert (1984). *The Sixties Papers: Documents of a Rebellious Decade*. New York: Praeger Publishers, 1984.

Arocena, Rodrigo & Sutz, Judith. "Latin American Universities: From an original revolution to an uncertain transition", *Higher Education* (2005), Universidad de la Repu´blica, Uruguay, pp. 573–592.

Blake, Bailey. *The 60s*. New York: Mallard Press, 1992.

Braungart, Richard G. and Braungart, Margaret M. "Youth Movements in the 1980s: A Global Perspective", *International Sociology*, Vol. 5, No. 2, June, 1990, pp. 155 to 181.

Cavan, Sherri. *Hippies of the Haight*. New Critics Press, St. Louis, MO, 1972.

Charnley, Gail. *Risk Analysis: Past, Present, and Future*. Presented at the International Conference on Risk Analysis and Its Role in the European Union. Belgium, 2000.

Coburn, Oliver. *Youth Hostel Story*. National Council of Social Service, 1950.

Davis, Joshua Clark. "The business of getting high: head shops, countercultural capitalism, and the marijuana legalization movement." The Sixties 8, no. 1 (2015): 27–49.

Gitlin, Todd. *The Sixties: Years of Hope, Days of Rage*. Toronto: Bantam Books, 1987.

Hoffman, Abbie. *Revolution For the Hell of It*. New York, NY: Thunder's Mouth Press, 1968.

Jamieson, Lee. *Antonin Artaud: From Theory to Practice*. London: Greenwich Exchange, 1970.

Katsiaficas, George N. *The Imagination of the New Left: A Global Analysis of 1968*. Boston: South End Press, 1987.

Kern, Edward. "Can It Happen Here?", *LIFE Magazine*, October 17, 1969.

Kitman, Jamie Lincoln. "The secret history of lead." *NATION-NEW YORK-* 270.11 (2000): 11.

Kovarik, William. *Chronology of Lead and Unleaded Gasoline*, Radford University, November 18, 2011. Accessed: March 29, 2022. Link: https://environmentalhistory.org/20th-century/roaring-1920s.

Krassner, Paul. *Confessions of a Raving, Unconfined Nut: Misadventures in the Counter-culture*. Simon & Schuster, 1994.

Lytle, Mark Hamilton. *The Gentle Subversive: Rachel Carson, Silent Spring, and the Rise of the Environmental Movement*. Oxford University Press, 2007.

Macleod, David I. *Building Character in the American Boy*. Madison, WI: University of Wisconsin Press, 1983.

Macleod, David I. *Building Character in the American Boy: The Boy Scouts, YMCA, and Their Forerunners, 1870-1920*. Madison, WI: University of Wisconsin Press, 2004.

Mandelbaum, Michael. "Vietnam: The Televised War", *Daedalus*, Vol. 111, No. 4, Print Culture and Video Culture (Fall, 1982), pp. 157–169.

Mazzini, G. *Manifesto del comitato centrale democratico Europeo agli Italiani* [1851], Scritti Editi ed Inediti di Giuseppe Mazzin, Vol. 46. Imola: Paolo Galeati, 1926.

Mazzini, Guiseppe. *Joseph Mazzini: His Life, Writings, and Political Principles.* New York: Hurd and Houghton, 1872, pp. 71–74.

McLuhan, Marshall. *The Gutenberg Galaxy: The Making of Typographic Man.* Toronto, Canada: University of Toronto Press, 1962.

Mettler, Katie. "The day Anti-Vietnam War Protesters Tried to Levitate the Pentagon". *Washington Post*, October 19, 2017. Accessed: April 12, 2022. Link: https://www.washingtonpost.com/news/retropolis/wp/2017/10/19/the-day-anti-vietnam-war-protesters-tried-to-levitate-the-pentagon.

Milanesio, Natalia. "Gender and Generation: The University Reform Movement in Argentina, 1918", *Journal of Social History*, Winter, 2005, Vol. 39, No. 2, *Kith and Kin: Interpersonal Relationships and Cultural Practices* (Winter, 2005), Oxford University Press, pp. 505–529.

Murdock, Barbara Scott. "Tracking Toxins: Biomonitoring Outshines the Indirect Assessment of Exposure in Determining Which Pollutants Enter the Body, and Whether They Cause Disease or Disability", *European Molecular Biology Organization (EMBO) Report*, August 2005; 6(8): 701–705. Accessed: March 29, 2022. Link: http://www.ncbi.nlm.nih.gov/pmc/articles/PMC1369152.

Murray, Lorraine, ed. *Italy.* New York: Brittanica Educational Publishing, 2014.

Nagy, Gabriella. "The Nexus between Hiking and Youth Hostels (1907–1933): A Historical Analysis of the Evolution of the German Youth Hostel Movement." *Journal of Outdoor Recreation and Tourism* 23 (2018): 59–66. https://doi.org/10.1016/J.JORT.2018.07.006.

O'Brien, Elizabeth and UNEP Partner. "Chronology of leaded gasoline/leaded petrol history." *The Lead Group Inc* (2011): 1–20. Accessed: March 29, 2022. Link: https://lead.org.au/Chronology-Making_Leaded_Petrol_History.pdf

Patterson, Clair C. "Contaminated and Natural Lead Environments of Man", *Archives of Environmental Health: An International Journal*, 11:3, 344–360, 1965. https://doi.org/10.1080/00039896.1965.10664229.

Robertson, Priscilla. "Students on the Barricades: Germany and Austria, 1848", *Political Science Quarterly*, Vol. 84, No. 2 (Jun., 1969), pp. 367–379.

Saad, Lydia. "Socialism as Popular as Capitalism Among Young Adults in U.S.", *Gallup.* November 25, 2019.

Sanders, Sam. "Why Do Young People Like Socialism More Than Older People Do?", *NPR*, September 21, 2015.

Savage, Jon (2008). *Teenage: The Prehistory of Youth Culture: 1875-1945.* London: Penguin Books, 576 pp.

Stark, Gary D. "The Ideology of the German Burschenschaft Generation", *European Studies Review* (SAGE, London and Beverly Hills), Vol. 8 (1978), 323–48.

Taylor, David. & Morris, Sam. (August 19, 2018). The Whole World is Watching: How the 1968 Chicago 'Police Riot' Shocked America and Divided the Nation", *The Guardian*. Accessed: April 20, 2022. Link: https://www.the-guardian.com/us-news/ng-interactive/2018/aug/19/the-whole-world-is-watching-chicago-police-riot-vietnam-war-regan.

Walker, Daniel. *Rights in Conflict: The Violent Confrontation of Demonstrators and Police in the Parks and Streets of Chicago during the Week of the Democratic National Convention of 1968*. Vol. 3852. New American Library, 1968.

Walker, Gabrielle. *An Ocean of Air: Why the Wind Blows and Other Mysteries of the Atmosphere*, Bloomsbury, 2007.

Walter, Richard J. "The Intellectual Background of the 1918 University Reform in Argentina", *Hispanic American Historical Review*, 1 May 1969; 49 (2): 233–253.

Williams, John Alexander. "Ecstasies of the Young: Sexuality, the Youth Movement, and Moral Panic in Germany on the Eve of the First World War", *Central European History*, 2001, Vol. 34, No. 2 (2001).

Zwicker, L. F. (2009). The Burschenschaft and German Political Culture, 1890–1914. *Central European History*, *42*(3), 389–428. https://doi.org/10.1017/S0008938909990033.

CHAPTER 2

# An Inconvenient Youth

> When you have so many people listening to what you're saying and especially when you write speeches that will be heard by a lot of people, you need to have solid ground, you can't make assumptions, you need to have facts on everything and sources on everything.
>
> Greta Thunberg
> Speaking on her process of speech preparation (McCarthy 2019).

As introduced in the previous chapter, the Groundswell Approach has seen varying degrees of achieving the goals of the youth engaged in creating causes in large groups. Often centric in nature or national in scope, there has rarely been a cause embraced by youth that is truly global and not just international until climate change. The obvious reason is that climate change predictions paint a bleak picture for the environment that today's youth will inherit as adults. They have a vested interest in stopping and reversing the scientifically identified trend towards global warming and the resulting devastating climate events. The not-so-obvious reason is that climate change is the first global issue that unites all youth, regardless of nationality and location of residence, and that they are able to mobilize themselves and their protests like never before due to the global village established by the Internet and the ubiquitous interconnectivity of its social media and associated hardware especially mobile devices.



M. Terry, *Speaking Youth to Power*, Palgrave Studies in Media and Environmental Communication,
https://doi.org/10.1007/978-3-031-14298-7_2

Encouraged by previous "wins" against business and industry in banning lead in gasoline, dichloro-diphenyl-trichloroethanes (DDTs), and chlorofluorocarbons (CFCs) to prevent death and destruction to the environment and its people, young people in the twenty-first century believe they, too, can stem the tide and reverse course ending the threat imposed by climate change before it becomes too late. They also recognize the urgency of their demands and the importance of being heard as opposed to merely being listened to.

## 2.1 Eco-Kids

One of the first mobilizations of youth in large cross-border numbers can be traced back to a US educational program called Eco-Kids in 1976. As other civil rights movements for women and peace were gaining momentum in the USA, the emerging environmental movement was focused on the issue of the day: air pollution. To capitalize on the growing interest in this issue, educator Loretta Bell created an education program called *Eco-Kids: Experiment with Air on Spaceship Earth*. The manual for the program was sponsored by the Bureau of Elementary and Secondary Education and published by the US Department of Health, Education, and Welfare and the National Institute of Education providing national endorsement and support of a program originally aimed only at the Highline Public Schools of Seattle, Washington.

Designed for children in elementary intermediate grades, the program describes Eco-Kids as "champions of what is good and clean and are friends of planets and animals. They are aware of all pollutions and are constantly battling these mysterious enemies for survival of Spaceship Earth" (Bell 1976). As the definition suggests, the program encouraged students to take a more active role in "battling" this environmental issue and suggested a global focus with its playful reference to "Spaceship Earth", a term first introduced by R. Buckminster Fuller in 1969.[1] It was an apt metaphor at a time when man first landed on the moon and space travel captured the imagination of everyone, especially children. He used it to suggest the ecological concept that humans are passengers on a vessel and that this ship, if damaged, affects the lives of all who travel in it. His

[1] Fuller's first reference to the term "Spaceship Earth" can be traced back to 1950 in his lectures at MIT, Black Mountain College, Harvard University, and other academic venues (Fuller 2001).

book was titled *An Operating Manual for Spaceship Earth* and provided an engaging context for understanding the impacts of global environmental issues like air pollution. In the book, Fuller suggests that younger minds may be better at finding solutions to these problems than adult minds.

> Because our spontaneous initiative has been frustrated, too often inadvertently, in earliest childhood we do not tend, customarily, to dare to think competently regarding our potentials. We find it socially easier to go on with our narrow, shortsighted specializations and leave it to others – primarily to the politicians – to find some way of resolving our common dilemmas. Countering that spontaneous grownup trend to narrowness I will do my, hopefully 'childish' best to confront as many of our problems as possible by employing the longest-distance thinking of which I am capable (Fuller, 21–22).

The connection made by Fuller between youth and the science-fictional metaphor of Earth as a spaceship resonated with educators like Bell and served as an effective launchpad to the Eco-Kids youth movement a decade later. On December 24, 1990, *TIME* magazine published a provocative report on the growing Eco-Kids movement, now in full swing across the USA and emerging in other countries. Perhaps echoing the sentiment of its readers at the time, the subtitle to the article "The Ecokid Corps" read, in part, "School-age crusaders can be a pain in the neck". While highlighting their environmental efforts, the report at times takes a dismissive tone referring to the young activists as "kiddies", "youngsters", and "do-gooders" (Elmer-Dewitt, 51). The article describes Eco-Kids as "the new generation of conservation-conscious, environmentally active schoolchildren" and suggests the movement is a result of media influence:

> Bombarded with ecomessages at school, in the press, on TV, and in pop-music lyrics, the youngsters have become convinced that they were put on the planet for the express purpose of saving it (Elmer-Dewitt, 51).

The bias throughout the article is that this young generation have been somehow brainwashed by disruptive adults who were "raised during the activist 1960s" adding that these activist adults are "eager to mobilize children (and) do not hesitate to use show-biz, though some might call it propaganda"—the author does not identify who those accusers might be. The inherent assumption of the author is that the youth of the 1990s would not have any inclination to care about nature had they not been raised by hippies and yippies, now assuming the grownup roles of parents, educators, and media creators. The report also acknowledges the successes of the movement, highlighting its global appeal in countries outside the USA such as Sweden, Japan, Costa Rica, and in Brazil where the article reports that the Eco-Kids movement has grown from 500 to 4000 in only three years (Elmer-Dewitt, 51). While versions of the Eco-Kids movement sprung up in these and other countries, there was not a cohesive mandate for its participants. Spurred on by the former hippie and yippie parents and teachers of this new generation, the Eco-Kids movement was unique to each country. While the general environmental theme of industrial pollution prevailed in each country, different issues specific to each nation became the focus. At this emerging stage, the Groundswell Approach to the youth environmental movement was global in theme, international in structure, but national in practice.

Perhaps most significant with respect to this early incarnation of environmental youth activism is the story of a sixth-grade class in Salt Lake City who collectively altered the Environmental Protection Agency of a hazardous waste dump with more than 50,000 barrels of crude oil which they promptly cleaned up. The government agency followed-up this action with a bill to pay for such clean ups in the future. This is the beginning of the US law that holds polluters accountable for illegal dumping and environmental damage.

This early victory achieved by a relatively small gathering of youth established evidence that you can indeed fight City Hall and win, an inspirational and motivational message for youth activism and environmentalists alike. The example set by Rachel Carson was still fresh in the minds of many as another example of how the individual can indeed impact on the policies of government and industry when addressing environmental issues that affect everyone. Apolitical in nature, environmental issues, once identified as a threat by the scientific community, were causes everyone could support, everyone except those who were causing them, of course.

In a follow-up story 30 years later, *TIME* magazine caught up with one of the young activists profiled in the 1990 article, Elizabeth Bayley who was 17 at the time. In 2019 she told *TIME* that she felt like an "enviro-geek" and an "outsider" in those early days, but sees that today's generation of youth feel quite the opposite. Today, "promoting environmental causes is more socially acceptable", she says, and attributes a lot of that coolness to the celebrity leadership of such movie stars as Leonardo DiCaprio and Mark Ruffalo (Waxman 2019).

Another youth profiled in the 1990 article, Kimberley Carr (now Kimberly Bryant), believes the groundswell of youth environmental activism today has a lot to do with telecommunications technologies. "Because of the Internet, there's so much that kids can get involved in now", she tells *TIME* in a phone call from Belize. "It's so much easier to be involved than it was when I was a kid" (Waxman 2019).

## 2.2 The Montreal Protocol and the Ozone Hole

One of the biggest wins for environmentalists of global citizens alike at this time that served to inspire the current youth climate activism, was the creation of the Montreal Protocol on Substances that Deplete the Ozone Layer established by the United Nations on September 15, 1987. To this day, it is the only UN treaty ever that has been ratified by every country on Earth—all 198 UN Member States (UNEP, n.d.). It was the scientific community, not youth environmental activists that informed and influenced the policymakers in Montreal, but like the Salt Lake City incident, the unilateral agreement provided evidence that environmental issues—even the ones on a global scale—can be addressed by governments and industry in a progressive legislative manner that protects the planet. The Montreal Protocol bans the ozone-depleting substances (ODS) chlorofluorocarbons (CFCs) and hydrochlorofluorocarbons (HCFCs), gases found in aerosol sprays, air conditioning units, blowing agents for foams and packing materials, as solvents, and as refrigerants. These gases are not only ozone-depleting substances, but also greenhouse gases that are 2000 times more potent than carbon dioxide in terms of its global warming potential (UNEP, n.d.).

A few years following the ratification and enforcement of the Montreal Protocol a disgruntled global industrial community nevertheless found ways of adapting the manufacture of their products without the harmful ODS.

> Makers of foam boxes for hamburgers and other fast foods now use materials with no ozone-ruining agents. Electronics firms and other high-tech manufacturers eliminated CFC-based solvents, switching to cleaning processes that rely on water or citrus compounds. Rocketdyne, the Canoga Park-based aerospace and defense contractor that formerly used CFCs to wash metal parts, now employs a water-vibration technique used for years in ultrasonic denture cleaners. Some manufacturers say they have actually saved

> money by dumping CFCs, which require more steps in the cleaning process. As industry has begun to wean itself from the chemicals, domestic CFC production has plummeted. According to the U.S. Environmental Protection Agency, about 126,000 tons of CFCs were manufactured last year, a 62% drop from 1986. Nonetheless, ridding American industry of CFCs remains a monumental task, especially in the air-conditioning and refrigeration sectors (Cheevers 1994).

While the Eco-Kids and other youth environmental groups were not targeting the ozone hole issue nor were they influential in its subsequent resolution in Montreal, the landmark environmental agreement signalled some very important procedures and partners to young environmentalists: the United Nations, as a legislator for global environmental issues; industry, as the frequent architect of environmental damage; and the scientific community, as the reliable harbingers of the reasons for environmental decline and of the actions needed to be taken to counter them. The Montreal Protocol also provided an example of the successful interaction of all three parties providing hope and inspiration to young people (and everyone else) that any environmental issue, not just on a local or national level, but on a global scale, can actually be resolved.

For the growing environmental youth movement, they began to see their place in this framework. Usually not a factor in discussions involving local politics, global policy, and business practices due to their age and relative lack of knowledge and experience, young environmentalists entering the twenty-first century knew they were armed with digital communication tools that could mobilize millions and amplify their voices to these key stakeholders. All they were lacking now was a leader.

## 2.3 Greta Thunberg

Arguably, the face of youth climate activism throughout the world is Greta Tintin Eleonora Ernman Thunberg, the now 18-year-old Swedish activist who began her campaign to end climate change at the age of 15 when she introduced her Skolstrejk för Klimatet (School Strike for Climate) outside the Swedish parliament on August 20, 2018.

She was motivated to do this as a means of battling depression caused by learning about the climate crisis in school when she was only eight years old. In a 2019 interview with *The Guardian*, she explained what happened:

> I remember when I was younger, and in school, our teachers showed us films of plastic in the ocean, starving polar bears and so on. I cried through all the movies. My classmates were concerned when they watched the film, but when it stopped, they started thinking about other things. I couldn't do that. Those pictures were stuck in my head (*Watts* 2019).

This led to a crippling depression that kept her from returning to school. She discussed her concerns with her parents who were consoling but dismissive of the actual direct impact of climate change on them. Undeterred by their relative lack of interest in the issue, she bombarded them with pictures, films, reports, graphs, and articles. Eventually, she got through to them. "That's when I kind of realised I could make a difference. And how I got out of that depression was that I thought: it is just a waste of time feeling this way because I can do so much good with my life" (*Watts* 2019). Seeing the change in her parents mindset made her think she may be able to similarly alter the mindsets of others. Following the news of forest fires in the Swedish Arctic, Thunberg decided she had to do something and went to her country's Parliament buildings for her now-famous solitary strike from school.

> I painted the sign on a piece of wood and, for the flyers, wrote down some facts I thought everyone should know. And then I took my bike to the parliament and just sat there," she recalls. "The first day, I sat alone from about 8.30 am to 3 pm – the regular schoolday. And then on the second day, people started joining me. After that, there were people there all the time. (*Watts* 2019)

Her modest one-person protest at the time turned into a global phenomenon known as Fridays for Future (FFF) beginning with more than 20,000 students walking out of classes around the world just a few months after her appearance outside the parliament buildings holding her iconic handmade sign. But this groundswell reaction would not have been possible in another time.

Although some have suggested Greta Thunberg is being manipulated by her parents (Wright 2019) and other adults (SkyNews 2019), there is no evidence to suggest she is and that the content in her social media accounts are indeed her own. The fact that she has enormous numbers of followers suggests she is the leader that this generation of youth have been waiting for. Her courage in speaking climate truths to the media, UN policymakers, and world leaders alike is inspiring to many, but especially youth in general, and those with autism (Thunberg has been diagnosed with Asperger's syndrome and is on the autistic spectrum (Hartwell et al. 2021)) who see her as a role model to counter injustice when perceived in authority, to question what has been considered normal practice for years, and to demand change in spite of being too young to vote (Lepkowska 2019).

Her relatability to her followers with respect to age is obvious and clearly a motivating factor of her appeal to youth today; however, peer relatability is not enough for an individual to command such a global following. Her candour and passion evident in her objectively moral position on an equally objective global environmental crisis are key components comprising *Transformational Leadership Theory* (TLT), a concept first identified in 1978 by James MacGregor Burns and later refined by Bernard M. Bass in 1985. Burns defines transformational leadership as when "leaders and followers make each other to advance to a higher level of morality and motivation" (Burns 1978). Through their unique personality and convictions, transformational leaders are able to change people's previously accepted perceptions and motivate them to demand change in the systems and structures that currently support or enable the status quo. Thunberg's uncompromising devotion to dismantling the Capitalistic systems that have helped define the age of the Anthropocene and that have contributed directly to the global changes in the climate define her as one such transformational leader.

## 2.4 The Four Elements of a Transformational Leader

Using a business structure as a model, the expanded version of TLT developed by Bass introduces four elements that define a transformational leader, all of which can be seen to apply to Thunberg:

### *2.4.1 Idealized Influence*

Transformational leaders act as role models and display a charismatic personality that influences others to want to become more like the leader. Idealized influence can be most expressed through a transformational leader's willingness to take risks and follow a core set of values, convictions, and ethical principles in the actions (they) take. It is through this concept of idealized influence that the leader builds trust with (their) followers and the followers, in turn, develop confidence in their leader.

### 2.4.2 *Inspirational Motivation*

Inspirational motivation refers to the leader's ability to inspire confidence, motivation, and a sense of purpose in (their) followers. The transformational leader must articulate a clear vision for the future, communicate expectations of the group, and demonstrate a commitment to the goals that have been laid out. This aspect of transformational leadership requires superb communication skills as the leader must convey (their) messages with precision, power, and a sense of authority. Other important behaviours of the leader include (their) continued optimism, enthusiasm, and ability to point out the positive.

### 2.4.3 *Intellectual Stimulation*

Transformational leadership values creativity and autonomy among the leader's followers. The leader supports (their) followers by involving them in the decision-making process and stimulating their efforts to be as creative and innovative as possible to identify solutions. To this end, the transformational leader challenges assumptions and solicits ideas from followers without criticizing. She helps change the way followers think about and frame problems and obstacles. The vision the leader conveys helps followers see the big picture and succeed in their efforts.

### 2.4.4 *Individualized Consideration*

Each follower or group member has specific needs and desires. For example, some are motivated by money while others by change and excitement. The individualized consideration element of transformational leadership recognizes these needs. The leader must be able to recognize or determine—through eavesdropping or observation—what motivates each individual. Through one-on-one coaching and mentoring, the transformational leader provides opportunities for customized training sessions for each team member. These activities allow team members to grow and become fulfilled in their positions. Each element is connected because there is a basis of respect, encouragement, and influence that is involved in transformational leadership (Lut and Lazoc 2020).

In Thunberg's case, she demonstrates her *Idealized Influence* by taking risks. In 2018, at the age of 15, she was the first—and, at the time, only—student to "strike" from school to sit outside Sweden's Parliament in a solo protest to pressure the government to meet carbon emissions targets (BBC 2021).

This initial act of defiance provided the *Inspirational Motivation* to attract followers not only in her home country, but across the globe. Today, her "Strike for Climate" marches are attended by thousands of youth often with replicas of her own original handmade sign. Despite her often dire messages of planetary destruction, she continues to display hope and optimism:

> I often talk to people who say, 'No, we have to be hopeful and to inspire each other, and we can't tell [people] too many negative things' ... But, no — we have to tell it like it is. Because if there are no positive things to tell, then what should we do, should we spread false hope? We can't do that, we have to tell the truth. (Thunberg, 2019)

She provides *Intellectual Stimulation* to her followers and those in power whom she confronts with irrefutable scientific data and by challenging the generally accepted assumption that constant meetings will eventually resolve the problem. Her infamous "Blah, blah, blah" (*The Guardian*, 2021) speech tapped in to a universal sentiment that political discourse is often ineffective.

Lastly, the way by which Thunberg displays *Individualized Consideration* is by understanding that her followers, primarily young people like herself, want a safe, secure, and healthy future. The common goal of all young people is not to inherit a world destroyed by their predecessors. This desire is frequently echoed by today's youth climate activist and represents a collective individual mindset of Thunberg's generation.

Critics of the Transformational Leadership Theory believe it needs to be further expanded to include direct expertise with the subject of desired change and the ability to provide viable solutions that can effectively be incorporated into the systems that impact on the subject. They call this *Instrumental Leadership Theory* (Antonakis and House 2014). In their essay on the subject they propose that successful leaders require these additional attributes:

> Effective leaders must also ensure that organizations adapt to the external environment and use resources efficiently (Fleishman et al., 1991; Hackman & Wageman, 2005; Morgeson, 2005; Mumford, 2006). That is, effective organizational leadership is not just about exercising influence on an inter-

> personal level; effective leadership also depends on leader expertise and on the formulation and implementation of solutions to complex social (and task-oriented) problems (Connelly et al., 2000; Mumford, Zaccaro, Harding, Jacobs, & Fleishman, 2000). Leaders must, inter alia, identify strategic and tactical goals while monitoring team outcomes and the environment (Morgeson, DeRue, & Karam, 2010). In this sense, and being true to the etymology of the word, leaders are "instrumental" for organizational effectiveness. (Antonakis and House 2014)

In Thunberg's case, the subject, of course, is climate change, and where she lacks the components of *Instrumental Leadership Theory* are in the areas of scientific and political expertise, and the ability to provide solutions that can be realistically incorporated into global systems and networks. Her age is usually pointed to as the reason she lacks these qualities and with them, many believe her efforts, while admirable and inspirational, fall short of what is necessary to achieve measurable impact.

Thunberg's undeniable fame is unprecedented in terms of numbers of followers for an environmental teenage leader (Instagram: 14.4 million;[2] Twitter: 5 million;[3] Facebook: 3.5 million;[4] TikTok: 346.4 million views;[5] YouTube: 911,333 views and 29.4 thousand followers[6]) making her a new kind of celebrity, an eco-star. Other notable politicians, movie stars, and pop music stars have similar numbers, but no environmentalist or eco-activist at any age in the world in history or today has ever achieved this level of measurable global fame. Despite lacking the credentials usually associated with environmental science (she has no earned university degree[7] or formal studies in the field) or government policy creation (she has no direct experience in policy writing or professional experience in politics) or business and industry (she has never engaged professionally in a Capitalistic business enterprise), Thunberg has nevertheless achieved leadership status on a global scale on the issue of climate change, itself an environmental issue on a global scale, primarily among the unique

[2] Instagram followers as of April 11, 2022: https://www.instagram.com/gretathunberg

[3] Twitter followers as of April 11, 2022: https://twitter.com/GretaThunberg

[4] Facebook followers as of April 11, 2022: https://www.facebook.com/gretathunbergsweden

[5] TikTok views as of April 11, 2022: https://www.tiktok.com/discover/Greta-Thunberg

[6] YouTube views and followers as of April 11, 2022: https://www.youtube.com/channel/UCAgIfWgzZ6QtvB_Oj1SBNnA/about

[7] She received an Honorary Degree from the University of British Columbia in 2021 in Canada and the University of Mons in Brussels in 2019.

demographic of youth, also on a global scale. Scholars questioning her authority in this privileged space, attribute her fame to a combination of a non-conformist brand of messaging and the support of conventional media bolstering her social media.

> ...through coverage of her international speeches and travel, statements by and relationships with other public figures, and her promotion of the voices of other climate change activists, Thunberg's combination of confrontational rhetoric, committed activism, lifestyles choices and public persona (e.g., the "ordinary school girl" of humble origins with Asperger's syndrome), connects her to some of the more foundational tropes and premises of environmentalism (apocalypticism, ecological jeremiad, importance of science). (Murphy, 194)

Dan Brockington, co-director of the Sheffield Institute for International Development at the University of Sheffield (UK), sees the phenomenon of Greta's undeserved authority as being representational. In particular, the public has an environmental consciousness that has changed over time and places. Celebrity environmentalism, he argues, "demonstrates the power of representations of nature and the authority of representation over experience" (Brockington, 553).

> The interaction between celebrity and environmentalism is not some convenient alliance between separate social phenomena. Rather the prominence of celebrity in environmentalism's affairs is a necessary consequence of the alienations of capitalism. (Brockington, 553–554)

Harkening back to a previous generation's suspicion of the "establishment" and the "man", Millennials are feeling alienated by the adults of capitalism who seemingly care more about profit than they do about the damage they may be causing to a planet today's youth will soon inherit. Those who already have a certain level of celebrity in the international entertainment industry (Leonardo DiCaprio, Cate Blanchett, Jackie Chan, etc.) are forgiven their lack of professional experience and related academic credentials in environmental issues since they speak for the concerns of youth and against those perceived to be causing climate change, a viewpoint widely shared by today's global community of youth, much the same way Rachel Carson spoke out against DDTs without being a scientist or a politician (see Chap. 1).

The communications tool that Carson did not have to bolster her eco-stardom was the Internet and the various social media platforms that amplify the reports of conventional media and other forms of non-fiction publishing. To demonstrate the power and reach of the Internet in the hands of even a young environmental activist, within only six months since Thunberg started her school strike by herself, she could command around 1.5 million youths in 100 countries to take the street for climate marches through her social media channels (Silberman 2019).[8]

The ubiquitous presence of social media platforms popular among youth worldwide has enabled Thunberg's messages and image to inspire young people on a global basis. The swiftness of the international growth of this movement reveals two things: that today's youth have not only a facility with digital communications tools, but a preference to engage with each other through them; and that today's youth are eager to demand a systemic change to what they see as their parents' generation's environmental negligence.

Research in this field suggests that the global communications technologies of social media embraced by today's youth are not used so much for protest recruitment as many might think, but rather to inform and influence each other about the subject of climate change, echoing the specific calls of the United Nations Framework Convention on Climate Change's (UNFCCC) Article 6. In a 2020 study conducted in Canada by MacEwan University (Edmonton) and Université du Québec (Trois-Rivières), mobilization of youth for participation in environmental protests such as Fridays for Future represents only 4.8% of posts on the Twitter platform. The other functions of tweets, the study reports, are for disseminating information on climate change (52.3%), opinion tweets (29%), and attack/blame tweets (13.6%). The study reports that in other social media, youth are engaging in a similar fashion: "information tweets (are) the most popular and mobilization the least popular".

Fridays for Future organizers are well aware of this trend in youth to better educate themselves about the science of climate change and the policies behind them. It appears that protest participation does not need to be encouraged or "mobilized". A well-informed and self-motivated generation of young people worldwide will eagerly attend these events on

[8] It should be noted that conventional media channels augmented her reach by picking up her social media stories, thereby amplifying her calls globally.

their own. The social media section of the Fridays for Future website offer a mosaic of the "newest (and) best posts of over 200 national FFF Social Media channels" most of which are tweets from Greta Thunberg, but many are "reports" of activities such as climate marches, climate conferences, and information on global climate data and policy. This reflects the results of the Canadian study that the youth of today are perhaps the most informed and networked global community of activists in history.

Climate change has become the cause célèbre of Millennials worldwide fuelled by their own social media platforms and fanned by extensive mainstream media coverage. Their protests are well-documented and their numbers impressive, but how effective are they in yielding the immediate and sweeping changes they demand. In his paper, *Fridays for Future's Disruptive Potential: An Inconvenient Youth Between Moderate and Radical Ideas*, Jens Marquardt casts doubt on their efficacy: "Although FFF has sparked debates about the urgency of tackling climate change, the movement's broader societal and political implications are yet to be seen". He suggests that the lack of results is related to a perceived threat by those in power that they may be replaced by the protestors and the radical ideologies they represent:

> A key motivator for social movements in general, and FFF activists in particular, is their high confidence in a future that is different from the established routines and the business-as-usual scenario. (Marquardt, 4)

This confidence is sparked by the earlier twentieth-century successes identified previously, but a lot of it also needs to be attributed to Thunberg. The world's youth see the teenager as a peer, not an untrustworthy adult who may have ulterior Capitalistic motives and not have their best interests at heart. Since she is a youth, her followers believe she is more accessible than other leaders. As influential as Thunberg was in the early days of this groundswell movement of youth environmentalism, there were a couple of significant predecessors in youth activism in the fight to end climate change that may have contributed to the groundswell approach among youth we now see on a global scale.

## 2.5 Severn Cullis-Suzuki

Severn Cullis-Suzuki, a Canadian environmentalist now 42, made an impassioned speech to the United Nations in 1992 when she was only 12 years old. Like Thunberg, Cullis-Suzuki's concern for the environment came early. At the age of 9, she founded the Environmental Children's Organization (ECO), a group of children dedicated to learning and teaching other youngsters about environmental issues. Reminiscent of the Eco-Kids of the 1970s, this organized group of youth of 12- and 13-year-olds was committed to making a difference and began a fund-raising campaign to send Cullis-Suzuki to the UN Earth Summit of 1992 in Rio de Janeiro where she made the following speech that rocked the world:

> Hello, I'm Severn Suzuki speaking for 'ECO' – the Environmental Children's Organization. We are a group of 12 and 13 year-olds trying to make a difference: Vanessa Suttie, Morgan Geisler, Michelle Quigg, and me. We've raised all the money to come here ourselves – to come 5,000 miles to tell you adults you must change your ways. Coming up here today, I have no hidden agenda. I am fighting for my future. Losing my future is not like losing an election, or a few points on the stock market.
>
> I am here to speak for all generations to come. I am here to speak – speak on behalf of the starving children around the world whose cries go unheard. I am here to speak for the countless animals dying across this planet, because they have nowhere left to go. I am afraid to go out in the sun now, because of the holes in our ozone. I am afraid to breathe the air, because I don't know what chemicals are in it.
>
> I used to go in – I used to go fishing in Vancouver, my home, with my Dad until, just a few years ago, we found the fish full of cancers. And now we hear of animals and plants going extinct every day, vanishing forever. In my life, I have dreamt of seeing the great herds of wild animals, jungles, and rainforests full of birds and butterflies, but now I wonder if they will even exist for my children to see.
>
> Did you have to worry of these things when you were my age? All this is happening before our eyes and yet we act as if we have all the time we want and all the solutions. I'm only a child and I don't have all the solutions. I don't – I want you to realize, neither do you. You don't know how to fix the holes in our ozone layer. You don't know how to bring the salmon back up in a dead stream. You don't know how to bring back an animal now extinct. And you can't bring back the forests that once grew where there is now a desert. If you don't know how to fix it, please stop breaking it. Here, you may be delegates of your governments, business people, organizers, report-

ers, or politicians. But, really, you are mothers and fathers, sisters and brothers, aunts and uncles – and all of you are someone's child.

I'm only a child, yet I know we are all part of a family – five billion strong; in fact 30 million species strong – and borders and governments will never change that. I'm only a child, yet I know we are all in this together and should act as one single world towards one single goal. In – In my anger, I'm not blind; and in my fear, I'm not afraid of telling the world how I feel. In my country we make so much waste, we buy and throw away, buy and throw away, buy and throw away and yet Northern countries will not share with the needy. Even when we have more than enough we are afraid to share; we are afraid to let go of some of our wealth.

In Canada, we live the privileged life. We've plenty of food, water, and shelter. We have watches, bicycles, computers, and television sets. The list could go on for two days. Two days ago, here in Brazil, we were shocked when we spent time with some children living on the streets. This is what one child told us: 'I wish I was rich and if I were, I would give all the street children food, clothes, medicines, shelter, and love and affection.'

If a child on the streets who has nothing is willing to share, why are we who have everything still so greedy? I can't stop thinking that these are children my own age, that it makes a tremendous difference where you are born; that I could be one of those children living in the favelas of Rio. I could be a child starving in Somalia, or a victim of war in the Middle East, or a beggar in India. I am only a child, yet I know if all the money spent on war was spent on finding environmental answers ending poverty and in finding treaties, what a wonderful place this Earth would be.

At school, even in kindergarten, you teach us how to behave in the world. You teach us to not to fight with others, to work things out, to respect others, to clean up our mess, not to hurt other creatures, to share, not be greedy. Then, why do you go out and do – do the things you tell us not to do? Do not forget why you are attending these conferences – who you're doing this for. We are your own children. You are deciding what kind of a world we are growing up in. Parents should be able to comfort their children by saying, 'Everything's going to be all right; it's not the end of the world, and we're – and we're doing the best we can.' But I don't think you can say that to us anymore. Are we even on your list of priorities?

My dad always says, 'You are what you do, not what you say.' Well, what you do makes me cry at night. You grown-ups say you love us. But I challenge you, please, make your actions reflect your words.

Thank you (Cullis-Suzuki 1992).

There are several similarities in the rhetorical style and content of this speech with those often given by Thunberg. Cullis-Suzuki begins by saying she is fighting for her future, a common theme in Thunberg's addresses and in the name of her protest movement, Fridays for Future. Both activists speak on behalf of their generation, their age group, with a forward vision common among climate researchers who often report predictive data. The two combine to provide an anticipatory vision of the years to come and our place in them.

Both activists also aim their criticism squarely on the adults in the room demanding they "change their ways". This particular point acknowledges that the current ways of doing things in government and in business are implicitly responsible for climate change, echoing descriptions and definitions of the Anthropocene, and as such, need to change to avoid global destruction. Both Thunberg and Cullis-Suzuki are making a big ask that would require systemic upheaval to business and government policies and frameworks. This becomes an important message as it suggests the myopic view to solving a problem is no longer effective. Global change as evidenced by the banning of CFCs and HFCs worldwide by the Montreal Protocol is an example of such radical change, only these young activists are calling for nothing short of a dismantling of capitalistic structures to rectify the global contributions to global warming and climate change industry causes. It is also interesting to note that Cullis-Suzuki's speech was made 26 years before Thunberg arrived on the scene, yet their messages—and passionate methods of delivery—are nearly identical suggesting a common methodology among today's youth that includes going straight to the top of global policy creation, the United Nations.

Cullis-Suzuki also addresses parental reaction to these concerns when children discuss the subject citing the common response "Everything's going to be all right", a dismissive comment intended to reassure children and alleviate their concerns. Nearly three decades later, Thunberg's own parents used this exact same line for the exact same reasons revealing that adults too, have not changed in their respective perspectives of the climate change issue.

While many point to Cullis-Suzuki as starting the global environmental youth movement in 1992 (Trihartono, Viartasiwi, Nisya), many others see the two young non-state actors as approaching their shared desire for change is very different ways: the Groundswell Approach and the Direct Approach, the primary focus of this book. Since her moving and persuasive speech to the UN Earth Summit in Rio de Janeiro, Cullis-Suzuki, instead of rallying the troops as Thunberg continues to do, shifted her focus to the source of global environmental policy change, the United

Nations. She is working closely with the UN on an organization called Earth Charter Commissioner that aims to continue raising the awareness towards climate change. In 2002, she was on the Special Advisory Panel for UN Secretary General at the 2002 UN World Summit on Sustainable Development in Johannesburg (Trihartono et al., 5). In this privileged capacity, she and members of the Skyfish Project brought their first policy proposal, a pledge called the "Recognition of Responsibility", to this conference. In a 2014 article with *Eye Magazine*, she recognizes her unique access to global environmental policymakers afforded her the opportunity to change the system from within to some degree, but 12 years later, the same access and opportunities to work directly with these policymakers has changed:

> When looking at the world today in comparison to 1992, many activists like Servern argue that the state of the environment is in a worse place, that globalization and corporate power has had devastating effect on the environment. Severn contends that the political landscape is very different now. 'We have world leaders that can't actually lead for the good of the people, I almost don't believe my words. When I see how beholden they are to corporations that have facilitated their rise to power in the current system, they are not able to make the leadership moves that will get us towards our carbon free economy, that will get us towards a world that will limit our global warming to under 2 degrees, so where does that leave us?... I think we have really ignored the strength of our democracies, we've really ignored our governmental processes – we have to strengthen that, we have to reform that'. (*Eye Magazine*, 27)

Both climate youth activists recognize, through distinctive experiences and engagements with politicians and policymakers at the UN level, that current government structures that reward political appointments through corporate donations need to change if effective climate policy is to be introduced worldwide.

## 2.6 Xiuhtezcatl Martinez

Another early youth climate activist "pre-Greta" is Xiuhtezcatl (pronounced shoe-tez-caht) Martinez, an American Indigenous hip hop performer from Colorado who co-founded Earth Guardians, "a global community of young leaders, activists, and artists defending our planet" (*Earth Guardians* 2022) in 1992.

On a smaller scale, Martinez practises the Groundswell Approach through art while at the same time engaging in the Direct Approach by addressing the United Nations and participating in the 2012 Rio+20 United Nations Conference on Sustainable Development in Rio de Janeiro when he was only 12. His calls to action at UN events recognize the importance of speaking directly to world leaders and UN policymakers, but falling short of direct meetings of productive collaboration, a challenge experienced by most youth climate activists. Perhaps his best-known encounter with power was in 2015 when he and 20 other youth activists sued the US government over its inaction to stop climate change. The lawsuit—*Juliana* versus *United States*—charges that the federal government has violated young people's constitutional rights to life, liberty, and property and this includes the right to a "stable climate system" (*Juliana*, "First Amended Complaint" 99).

In a 2020 interview with *Vogue Magazine*, Martinez explains the reasons for this unusual course of action:

> We filed a lawsuit against the Obama administration in 2015 because of its failure to adequately respond to the climate crisis—a crisis that scientists have known is a threat for decades now. Obviously in 2016, Trump took office and so our case pivoted to the administration in power. It is now in the hands of the Ninth Circuit court; as long as there's still an opportunity for us to have our voices heard, we're going to be fighting to get our trial. One of the coolest things about this lawsuit is that it's shown the world that it is legally possible to hold our governments accountable for climate inaction. (Chan 2020)

Martinez, like Thunberg in this regard, uses temporal arguments in his climate activism. Challenging such statements as "Our Children are Our Future", he invokes iterations of the past from his Indigenous culture. In his book, *We Rise: The Earth Guardians Guide to Building a Movement that Restores the Planet*, he emphasizes the present in the title and states "The future of the planet needs to be shaped by an important balance of technology, media, innovation, and indigenous wisdom," which would involve "Reconnecting to these ancient teachings" (*We Rise*, 12).

This approach has served Martinez and his Earth Guardians well in speaking youth to power, but he has not forgotten the importance of reaching out to youth and his Groundswell Approach to doing this is equally unusual. Martinez uses art (his book (*We Rise*), film (*Kid Warrior*

(2015)), and most notable "eco hip hop") to capture the attention of Millennials today. Eco hip hop is essentially hip hop music with environmental themes and messages. "I feel like my music sprouted from an understanding that speaking to an audience wasn't always gonna reach people. I started writing music as a way to just talk, using hip-hop to talk to young people" (Bulotano 2021).

Reminiscent of the protest songs of the 1960s, Martinez's music often directly calls out government and industry and rallies youth to take action. In his song Boombox Warfare, he sings:

> People singing songs of freedom
> while we marching [sic] in the streets.
> Yes this is critical. This is my resistance.
> Artists hold more power than these crooked politicians.
> This is Boombox Warfare. (Boombox 2019)

Connecting to youth through music is nothing new, but the messaging this time is the global environmental issue of climate change and since music is a universal language, it is appropriate to use this art form to reach the global community of youth. In her essay, *Singing Our Own Song: Navigating Identity Politics through Activism in Music*, Juliet Hess argues that "Activist-musicians provide a crucial perspective toward shaping music education towards activism and understanding music as an identity-making process. The hyphenated activist-musician identity offers a unique praxis-oriented identity. For many activist-musicians, their activist self is rooted in equity and critique, while their musician self allows them to assert these ideologies and critiques into the world" (Hess, 62).

On the other hand, while scholars such as Ambrose Leung and Cheryl Kier acknowledge the power of music in activist messaging and education, their research suggests that hip hop is not the best style of music to use: "A stream of literature also exists that categorizes hip-hop and rap as related to negative youth behaviour such as stealing, fighting, smoking tobacco, and using drugs…However, it may also be argued that such rebellious behaviour can be considered a form of resistance against social reality" (Leung, Kier, 446).

Thunberg's Groundswell Approach is the Fridays for Future marches organized globally through social media. Her messages centre around the science of climate change and are aimed at national and international governments, particularly Sweden and the United Nations. Martinez's

Groundswell Approach is through the creation of art products disseminated globally through digital artistic distribution platforms. His messages also focus on climate change and are aimed at national governments, particularly the United States. His messages—delivered directly to world leaders at the UN and by suing the US federal government—reach millions of youth in a related but different manner. His earth-based culture informs his messaging and demands and this aspect of his approach is quite different from that of Thunberg's. Both approaches reach millions around the world and their respective eco-celebrity, in part, has a lot to do with their very young age when they began their activism: Thunberg was 8;[9] Martinez was 9.[10] The extreme young age captured the imaginations of the public. As they grew older, their speeches and activist activities became more sophisticated as were their platforms (both have been invited to address the United Nations). As of this writing, however, neither activist can point to any substantial legislation, either national or international, that has been passed and implemented based directly on their influence which is their ultimate goal. Indirectly, there is no denying their respective Groundswell Approaches to reaching millions around the world and raising the awareness of a scientific issue like never before and this, perhaps, provides immeasurable impact on the policies that have been created.

## 2.7 Autumn Peltier

Another young climate activist leader with a high global profile and from an Indigenous community is Autumn Peltier from the Wiikwemkoong First Nation on Manitoulin Island, Ontario, Canada. She has held the title of Chief Water Commissioner for the Aniishnabek Nation since 2019. Her particular methodology in applying the Groundswell Approach is to focus on one climate-centric issue: water. Her rhetorical strategy not only focuses the attention on an aspect of climate change that everyone on the planet deals with directly on a daily basis, but also provides this perspective from an earth-based culture that has a more intimate, and therefore, more seemingly credible relationship to the global environment. Having grown up in an Indigenous community her relationship with the land and water is more personal than others living in urban environments. In 2018, and

[9] https://www.britannica.com/biography/Greta-Thunberg

[10] https://www.earthisland.org/journal/index.php/magazine/entry/xiuhtezcatl_martinez

at the age of 13, she told the General Assembly of the United Nations that "our water deserves to be treated as human, with human rights. We need to acknowledge our waters with personhood, so we can protect our waters…We all have a right to this water as we need it. Not just rich people, all people. No one should have to worry if the water is clean or if they will run out of water" (Kent 2018).

By personifying water with human attributes, including rights, Peltier adds a new dimension to the perception of water to the UN policymakers. This provides a more profound urgency to address water concerns as it now may be imbued with characteristics that legally bind the policymaker to take action. The ecological connection she implies is that since every human being needs water and cannot survive without it, water itself should be endowed with the same rights we give humans.

Like Thunberg and Martinez, Peltier also began her environmental activism at a very young age: eight. Inspired by the teachings of her great-aunt, Josephine Mandamin, from whom she inherited the role of Chief Water Commissioner for the Aniishnabek Nation, Peltier visited a community in Canada under a boiled water advisory and began further researching the issue and advocating for change alongside her elders (Korte 2019). "She's my hero. Ever since I was a little girl, she (Mandamin) taught me everything I needed to know about the importance of water... She was one of the most important people to me" (Korte 2019). Her great-aunt had travelled the shores of all five of the Great Lakes and alerted authorities of issues with the water when she found them. Following in these footsteps, Peltier also addresses water issues and brings them to the attention of authorities on a more global scale.

In a 2018 meeting with Canadian Prime Minister Justin Trudeau when she was only 12, she was able to voice her disappointment with his decision to allow the nationwide construction of oil pipelines: "She had been chosen to present the Canadian prime minister with a ceremonial copper water bowl to symbolise his responsibility to protect the country's water. But instead of being star struck, the teenager spoke her mind. 'I am very unhappy with the choices you've made,' she said. As prime minister, Mr. Trudeau has supported a number of pipeline projects, drawing sharp criticism from indigenous and environmental advocates. 'I understand that', he said to her. 'I will protect the water'" (BBC 2017).

But the water was not protected, as promised. In 2021, there were 59 active long-term water advisories in 41 communities across the country

(Romaniuk 2021) and a displeased Peltier spoke out once more this time accusing the prime minister and his government of racism.

> If water advisories like this were to happen in a place like Ottawa, just think about how fast the government would work to get clean drinking water again. It would probably be done in less than a week…Why is it that First Nations people have to wait so long for something so simple? Has it not been long enough?… "Water is a basic human right, and nobody should have to beg for it. This is wrong, and it's come to the point where I think it comes down to racism," said Autumn Peltier, a teenage water-rights activist from Wiikwemkoong Unceded Territory on Manitoulin Island. (Prime Minister) Justin Trudeau promised me personally that he was going to protect the water when I confronted him about his broken promises to my people. I would like to see the government stop making promises they can't keep. (Romaniuk 2021)

Like Martinez, Peltier can be seen here to use a hybrid of both the Groundswell and Direct Approaches, reaching many in her Indigenous communities and around the world through conventional media and social media coverage, but also appealing directly to world leaders individually (Trudeau) as well as collectively (the United Nations).

Another interesting manifestation of the Groundswell Approach is the snowball effect among youth climate leaders. While numbers of followers and climate activists are growing exponentially among the rank and file, new youth climate leaders like Martinez and Peltier are emerging inspired by the success and heroic presence of who is arguably the youth climate movement's founder, Greta Thunberg.

## 2.8 Influential Youth Climate Activists

According to its website, *Earth Day* is the "world's largest recruiter to the environmental movement", a non-profit organization with more than 50,000 partners in 192 countries. Its mission is to diversify, educate, and activate the environmental movement worldwide ("About Us", *earthday.org*). Serving as a central hub for youth climate activities, the organization recently identified who it believes represent the top youth activists in the world. This is a list of some of those new youth leaders along with their social media handles as reported on their website. The brief biographies detail some of their recent activities:

### 2.8.1 *Vic Barrett | USA*

> The journey of looking at myself and my identity… made it clear to me that something had to be done. How could you not when people are dying?

Vic is from low-lying land in New York and has felt first-hand climate impacts in the form of Hurricane Sandy. Vic is a Fellow with the Alliance for Climate Education, spoke at the COP21 UN Conference on Climate Change and spoke at the UN headquarters in New York City. After marching in solidarity with more than 400,000 people at the People's Climate March in New York City, he organized his peers in local frontline climate campaigns. Vic is now an undergraduate student at UW-Madison and is also among 21 youth activists who are actively suing the government to take action on climate change.

Follow Vic on Twitter @vict_barrett and Instagram vicbarrett_.

### 2.8.2 *John Paul Jose | India*

> India should declare a climate emergency. The rich biodiversity and culture are under direct threat from our climate crisis.

Twenty-two years old climate activist, writer, and global peace ambassador John Paul Jose provides a commentary on our environmental crisis through an Indian lens. From discussions on the impacts on India's iconic forests and trees to examinations on the connections between climate action and sustainable development, water and food security and more, John Paul literally covers a lot of ground in his climate activism in India. Now working with the High Seas Alliance, he has taken the visionary approach of gaining policy experience at the UN levels by working with several NGOs and UN agencies: the UN Major Group for Children and Youth (UNMGCY), the United Nations Environment Programme (UNEP), the United Nations Convention on Combatting Desertification (UNCCD), conducting research with The Energy Research Institute in India, and the TED Countdown Youth Council.

Follow John Paul on Twitter @johnpauljos and Instagram @johnpauljos.

### 2.8.3 *Luisa Neubauer | Germany*

> What the world community is doing with the planet, will be described one day as the biggest political failure of our time.

If you need a new female role model, meet Luisa Naubauer. At 23, poised and calm, she is talking the talk and walking the walk. She is a geography student, activist, and member of the German environmentalist Green Party, and in the last year, she has become the German face of the "Fridays for Future" climate protection rallies.

Follow Luisa on Twitter @Luisamneubauer and Instagram @ luisaneubauer.

### *2.8.4 Isra Hirsi | USA*

> It is my future, it is the younger ones' future, and it's less the older ones' future.

Isra Hirsi is the 16-year old co-founder and co-executive of the US Youth Climate Strike, the organization that leads the student climate strikes. Inspired by the Flint water crisis, Isra fights for climate justice. As a black Muslim woman, she emphasizes the importance of intersectionality in the movement for climate justice. She is also a part of MN Can't Wait, a youth coalition of organizations calling upon the Minnesota government to take action.

Follow Isra on Twitter @israhirsi and Instagram @israhirsi.

### *2.8.5 Holly Gillibrand | Scotland*

> When there's about 200 species going extinct every day, it's quite scary. So you have to do everything you can.

From her small town in the remote, mountainous Scottish Highlands, this 13-year old is helping to build a UK movement of children demanding more action on climate change. Holly Gillibrand is a young ambassador for Scotland: The Big Picture and a campaigner for animal welfare charity OneKind.

Follow Holly on Twitter @HollyWildChild

### *2.8.6 David Wicker | Italy*

> We will not stop until politicians and leaders decide to take action.

Fourteen-year-old David Wicker lives in Italy and has been organising groups of students to protest via Fridays for Future in Turin, Italy and internationally with the purpose of asking governments all around the world to place the Climate Change issue as their top priority in their agendas and to start respecting the regulation of international agreements and treaties.

Find David on Twitter @davidwicker_hf and Instagram @ davidwickerhf.

### *2.8.7 Lilly Platt | the Netherlands*

> You don't have to be a grown up to do something. Children are allowed to help the environment. If they don't, they won't have a future. They won't have anything to go to school for.

Just at nine years of age, Lilly Platt, started going on a war to end plastic pollution. When Lilly first moved to Holland her grandfather began teaching her to speak Dutch, and to help her learn numbers, they collected pieces of plastic trash and counted them. She no longer needs the practice but she's still picking up plastic. Today, Lilly is a Youth Ambassador for the Plastic Pollution Coalition and a Child Ambassador for HOW Global and World Cleanup Day.

Follow Lilly on Twitter @lillyspickup

### *2.8.8 Saoi O'Connor | Ireland*

> If a politician comes to my door, I start asking them hard questions. Some of them are shocked … The politicians we elect this weekend will have a direct influence over whether or not my generation inherits a liveable planet …

Every Friday morning, 16-year-old Saoi O'Connor travels two hours from her home to the city to protest. Saoi has been dissenting in her own way since the age of four when her parents started the Fair-Trade Committee in her town. Now she is one of the critical figures of Ireland's youth climate rebellion.

Follow Saoi on Twitter @saoi4climate.

### *2.8.9 Jamie Margolin | USA*

> We had no power in creating the systems that are destroying our world and futures — and yet we are and will be paying the biggest price for the older generations' recklessness.

Seventeen-year-old Jamie Margolin is the founder and co-executive director of *Zero Hour*, a youth- and women of colour-led movement that aims to support new young activists and organizers with the tools, training, and resources that they need to fight climate change and protect our natural resources. Jamie speaks passionately and urgently about the intersectional issues of social justice, environmental protections, climate change, indigenous rights, and more, and isn't afraid to call out corporate leaders and others who have contributed to our environmental crises.

Follow Jamie on Twitter @Jamie_Margolin and Instagram @ jamie_s_margolin.

### *2.8.10 Leah Namugerwa | Uganda*

> Most people do not care what they do to the environment. I noticed adults were not willing to offer leadership and I chose to volunteer myself. Environmental injustice is injustice to me.

As a 14-year-old member of Fridays for Future Uganda, Leah Namugerwa strikes every Friday for climate justice. Leah was inspired by Greta Thunberg but was truly moved to act after seeing the famine caused by drought and landslides from climate change. Despite the Ugandan government's harsh response to strikers, Leah continues to fight for change. Currently, she is demanding a Ugandan plastic bag ban.

Leah talked with us about what it's like to climate strike in Uganda, and how she ignored the critics who called her strikes "weird".

Follow Leah on Twitter @NamugerwaLeah.

### *2.8.11 Anuna De Wever | Belgium*

> This is what the youth wanted, a seat at the table. But of course nothing concrete has happened yet. Drastic measures must be taken to prevent global warming.

Inspired by Greta Thunberg, 17-year-old Anuna frequently strikes for climate action in Brussels. She believes that everyone has a role to play in combating climate change and is determined to convince politicians to respond appropriately to the climate crisis.

Follow Anuna on Twitter @AnunaDe and Instagram @anuna_dewever.

### 2.8.12 *Jerome Foster II | USA*

> Adults take note of this message: Young people like myself should not have to take on this burden, this is supposed to be your job but now we have to go on hunger strikes, meet with government officials, and start a global movement for you to even notice.

Jerome Foster II is no slouch—he's a climate activist, author, National Geographic Explorer, Smithsonian Ambassador, and Founder and Editor in Chief of *The Climate Reporter*. And did we mention that he's 16? Recently, Jerome hosted the Washington, DC youth climate strikes at the White House for the May 24 global strike movement, and spent the summer striking at Harvard University in Massachusetts.

Follow Jerome on Twitter @jeromefosterii.

### 2.8.13 *Eyal Weintraub | Argentina*

> We have reached a point in history when we have the technical capacities to solve poverty, malnutrition, inequality and of course global warming. The deciding factors for whether we take advantage of our potential will be our activism and our international unity.

Eighteen-year-old Eyal recently organized a protest in front of the national congress in Buenos Aires, with fellow activist Bruno Rodriguez. Eyal understands that his generation has the potential to solve this issue, but it will take cooperation and activism in order to make it happen. He is a member of Jóvenes Por El Clima Argentina, an organization of youth climate activists in Argentina.

Follow Eyal on Instagram @eyalwein.

### 2.8.14 *Alexandria Villaseñor | USA*

> You have to listen to the science and the facts because climate change isn't an opinion.

Alexandria is a 14-year-old activist who strikes outside of the United Nations in order to bring attention to climate change. She is an organizer of FridaysForFuture, an organization of students striking every Friday around the world. She is also the founder of Earth Uprising, a non-profit where youth around the world rise up against climate change together.

Follow Alexandria on Twitter @AlexandriaV2005 and Instagram @ alexandriav2005.

### *2.8.15 Haven Coleman | USA*

We can stop the worst effects, so why shouldn't I try and save all you adults?

At just 13, Haven Coleman is already a co-founder and co-executive director of the US Youth Climate Strike. While she has always been passionate about fighting against injustice, she has focused on climate change for the past several years. Despite the backlash she has received from classmates, politicians, and even strangers, Haven remains determined to continue her activism and bring about meaningful change.

Follow Haven on Twitter @havenruthie and Instagram @ climateactivist.

("19 Youth", 2019).

## 2.9 Conclusion

It is interesting to note that most of these youth climate activists are under the age of 18 while many of the United Nations Framework Convention on Climate Change's programs and activities for youth define this group as being between the ages of "18 to 30" (ACE 2017). This younger age for some of the world's most high-profile youth activists bodes well for their presence and participation in future UN climate summits when they become "of age". The experience they acquire today will give them the background in climate science and international policy their critics have traditionally accused them of lacking in the past. Their facility with digital technologies and global communications gives them an advantage over previous youth environmentalists who were not as well connected and interconnected. Power still resides in the hands of the government officials, policymakers, and negotiators of the United Nations, but today's youth climate activists are powering up themselves to better communicate, dialogue, and participate with the UN in forging progressive climate policies for themselves and their future generations.

The next chapter will examine the digital technologies, the social media, and the communications methodologies Greta Thunberg and her fellow youth climate leaders are using to mobilize their forces. Unlike previous generations of youth, today's youth have a global reach never before experienced affording them the power of instant communications and knowledge mobilization. These become important tools in the fight to be heard and to contribute directly to the world's climate policy process.

## Works Cited

"19 Youth Climate Activists You Should Be Following on Social Media", *earthday.org*, June 14, 2019. Accessed: April 19, 2022. Link: https://www.earthday.org/19-youth-climate-activists-you-should-follow-on-social-media

"ACE – Empowering youth and the public on taking climate action", *UNFCCC*, 2017. Accessed: May 30, 2022. Link: https://cop23.unfccc.int/news/ace-empowering-youth-and-the-public-on-taking-climate-action

BBC News. (2017, December 31). *Teen Activist Autumn Peltier Who Scolded Trudeau to Address UN*. Accessed: April 16, 2022. Link: https://www.bbc.com/news/world-us-canada-42358227

"'Blah, blah, blah': Greta Thunberg Lambasts Leaders over Climate Crisis", *The Guardian*. September 28, 2021. Accessed: April 10, 2022. Link: https://www.theguardian.com/environment/2021/sep/28/blah-greta-thunberg-leaders-climate-crisis-co2-emissions

Earth Guardians, *Instagram.com*. Accessed: April 12, 2022. Link: https://www.instagram.com/earthguardians/?hl=en

"Greta Thunberg is 'a miserable child manipulated by adults'", *SkyNews*, December 3, 2019. Accessed: March 30, 2022. Link: https://www.skynews.com.au/australia-news/greta-thunberg-looks-like-a-miserable-child/video/bc1cc0e2f8d23f650466dc2b2557950e

*Juliana, et al versus United States, et al*, US District Court for the District of Oregon, Case Number: 6:2015cv01517, filed August 12, 2015. Link: https://dockets.justia.com/docket/oregon/ordce/6:2015cv01517/123110

"Severn Cullis-Suzuki", *Eye Magazine*. Toronto: Issue 3, Spring, 2014.

UNEP (n.d.), *About Montreal Protocol*, UNEP.org. Accessed: April 9, 2022. Link: https://www.unep.org/ozonaction/who-we-are/about-montreal-protocol

"Xiuhtezcatl, feat. Jaden Smith – Boombox Warfare (Official Lyric Video)." YouTube, 26 April, 2019. Accessed: April 15, 2022. www.youtube.com/watch?v=5DA-w-_HZLo.

Antonakis, John and House, Robert J. "Instrumental Leadership: Measurement and Extension of Transformational–Transactional Leadership Theory", *The Leadership Quarterly*, 25, May 10, 2014, 746–771.

Bass, B. M. (1985). *Leadership and Performance beyond Expectation*. New York: Free Press.

Bell, Loretta. *Eco-Kids: Experiment with Air on Spaceship Earth*. Seattle, WA: Highline Public Schools and the Department of Elementary and Secondary Education, 1976.

Bulotano, Phoebe. “Xiuhtezcatl Martinez Inspires Youth with Eco Hip-Hop”, *ArtsHelp*, November, 2021. Accessed: April 15, 2022. Link: https://www.artshelp.net/xiuhtezcatl-martinez-eco-hip-hop/#:~:text=I%20started%20writing%20music%20as,advocate%20for%20the%20climate%20crisis

Burns, J.M. *Leadership*. New York: Harper & Row, 1978.

Chan, Emily. “Martinez on why it’s so important to support indigenous communities right now,” *Vogue Magazine*, July 3, 2020. Accessed: April 15, 2022. Link: https://www.vogue.in/culture-and-living/content/hip-hop-artist-and-activist-xiuhtezcatl-martinez-on-why-its-so-important-to-support-indigenous-communities-right-now

Cheevers, Jack. “Industry Shivers at Ban on CFC Refrigerants: Environment: An end to production of the ozone-eroding chemicals next year is forcing a costly transformation on American businesses”, *The Los Angeles Times*, November 1 1994. Accessed: April 9, 2022. Link: https://www.latimes.com/archives/la-xpm-1994-11-01-fi-57325-story.html

Cullis-Suzuki, Severn. “Speech at U.N. Conference on Environment and Development”, *American Rhetoric*, 1992. Accessed: April 12, 2022. Link: https://www.americanrhetoric.com/speeches/severnsuzukiunearthsummit.htm

Elmer-Dewitt, Philip. “The Ecokid Corps”, *TIME Magazine*. New York: December 24, 1990.

Friedman, H., Langbert, M., (2000), “Abraham as a Transformational Leader”, *Journal of Leadership & Organizational Studies* Spring, vol. 7 no. 2, 88–95.

Fuller, R. Buckminster. *An Operating Manual for Spaceship Earth*. New York: Simon and Schuster, 1969.

Fuller, R. B. *Your Private Sky: Discourse* (J. Krause & C. Lichtenstein, Eds.). Baden, Germany: Lars Müller, 2001.

Hartwell, M., Keener, A., Coffey, S. *et al.* Brief Report: Public Awareness of Asperger Syndrome Following Greta Thunberg Appearances. *Journal of Autism and Developmental Disorders* (2021) 51:2104–2108. Link: https://doi.org/10.1007/s10803-020-04651-9

Hess, Juliet. “Singing Our Own Song: Navigating Identity Politics through Activism in Music”, *Research Studies in Music Education*, Vol. 41(1), 2019.

Hook, Leslie. “Greta Thunberg: ‘All my life I’ve been the invisible girl,'”, *Financial Times*, February 22, 2019. Accessed: May 5, 2022. Link: https://www.ft.com/content/4df1b9e6-34fb-11e9-bd3a-8b2a211d90d5

Kent, Melissa. “Canadian Teen Tells UN to ‘Warrior Up’ to Protect Water”, *CBC News*, March 22, 2018. Accessed: April 16, 2022. Link: https://www.cbc.ca/news/canada/autumn-peltier-un-water-activist-united-nations-1.4584871

Korte, Kate. “Meet Autumn Peltier, the 15-year-old ‘Water Warrior’,” *The Martlet*, October 4, 2019. Accessed: April 16, 2022. Link: https://www.martlet.ca/meet-autumn-peltier-the-15-year-old-water-warrior

Kraemer, Daniel, "Greta Thunberg: Who is the climate campaigner and what are her aims?", *BBC News*, November 5, 2021. Accessed: April 10, 2022. Link: https://www.bbc.com/news/world-europe-49918719

Lepkowska, Dorothy. "A Role Model for All Young People Living with Asperger", *British Journal of School Nursing*, Vol. 14, No. 4, May 17, 2019. Link: https://doi.org/10.12968/bjsn.2019.14.4.198

Leung, Ambrose and Kier, Cheryl. "Music preferences and civic activism of young people", *Journal of Youth Studies*, 11:4, 445–460, 2008.

Lut, Dina and Lazoc, Alina. "The Impact of Transformational Leadership on Organizational Effectiveness", *ResearchGate*, researchgate.net. May 9, 2020. Accessed: April 10, 2022. Link: https://www.researchgate.net/publication/341267219_THE_IMPACT_OF_TRANSFORMATIONAL_LEADERSHIP_ON_ORGANIZATIONAL_EFFECTIVENESS

McCarthy, Joe. "Greta Thunberg: 10 Quotes From The Climate Activist", *Global Citizen*. Globalcitizen.org, September 10, 2019. Accessed: March 29, 2022. Link: https://www.globalcitizen.org/en/content/greta-thunberg-quotes-naomi-klein

Marquardt, Jens. "Fridays for Future's Disruptive Potential: An Inconvenient Youth Between Moderate and Radical Ideas", *Frontiers in Communication*, 22 July 2020. Accessed: December 7, 2021. Link: https://doi.org/10.3389/fcomm.2020.00048

Martinez, Xiuhtezcatl. *We Rise: The Earth Guardians Guide to Building a Movement that Restores the Planet*. Rodale, 2017.

Murphy, Patrick D. "Speaking for the Youth, Speaking for the Planet: Greta Thunberg and the Representational Politics of Eco-celebrity", *Popular Communication*, 19:3, 2021, pp. 193–206.

Romaniuk, Colleen. "Clean Water for First Nations Critical During Pandemic: Activists", *Sudbury Star*, January 6, 2021. Accessed April 16, 2022. Link: https://www.thesudburystar.com/news/local-news/clean-water-for-first-nations-critical-during-pandemic-activists

Silberman S. "Greta Thunberg 2019 became a climate activist not in spite of her autism, but because of it", *Vox* [Internet], September 24, 2019. Accessed: April 12, 2022. Link: https://www.vox.com/first-person/2019/5/6/18531551/greta-thunberg-autism-aspergers

Trihartono, A. et al. "The giant step of tiny toes: youth impact on the securitization of climate change", East Java, Indonesia: 2020 IOP Conference Series: *Earth and Environmental Science*, Vol. 485, September 28–29, 2019.

Watts, Jonathan. "Greta Thunberg, schoolgirl climate change warrior: 'Some people can let things go. I can't'," *The Guardian*, March 11, 2019. Accessed: April 12, 2022. Link: https://www.theguardian.com/world/2019/mar/11/greta-thunberg-schoolgirl-climate-change-warrior-some-people-can-let-things-go-i-cant

Waxman, Olivia B. "'I Don't See the World With Rose-Tinted Glasses.' What Happens When Young Environmental Activists Grow Up", *TIME Magazine*. New York: March 15, 2019. Accessed: April 8, 2022. Link: https://time.com/5549331/youth-climate-strike-ecokids

Wright, Jennifer. "These Anti-Greta Thunberg Criticisms Are Almost Hilarious", *Harper's Bazaar*. New York: Hearst Magazines Digital Media, September 26, 2019.

CHAPTER 3

# Opening the Door to Negotiation

(The protest) seeks to create such a crisis and foster such a tension that a community which has constantly refused to negotiate is forced to confront the issue. It seeks so to dramatize the issue that it can no longer be ignored. Martin Luther King's essay, *The Letter from Birmingham Jail.*

After examining the styles of contemporary youth leadership in effecting their respective Groundswell Approaches, this chapter will examine the methodologies they use to grow and command their numbers. Many of these means are relatively new to youth activism having not existed prior to the 1990s. The digital technologies of the Internet and personal devices have not only made the world smaller as Marshall McLuhan predicted in 1962 when he coined the term "the global village", but size, cost, and accessibility have provided instant communication with youth the world over for leaders and followers alike. Gone are the days of mimeographing posters and pasting them to walls to advertise rallies of the 1960s; gone also are the days of relying on the conventional media of radio and television to spread the word of the causes of youth in the 1970s and 1980s. Today's youth need simply to refer to their phones and open their social media platforms to read the latest message about protest marches near them, often accompanied by geomedia support such as Geographic





M. Terry, *Speaking Youth to Power*, Palgrave Studies in Media and Environmental Communication,
https://doi.org/10.1007/978-3-031-14298-7_3

Information Systems (GIS) and Global Positioning Systems (GPS) that can direct them to the event in real time. Another significant feature of this ubiquitous digital media is its ability to provide youth with the visible evidence necessary to illustrate the impacts of climate change through photography and video and disseminate these images to those in power leaving little doubt in the minds of today's youth and environmental policymakers of the importance of climate change action.

The cause of climate change is both one of global concern—not just to youth, but to everyone on the planet—as well as an issue that is apparently so big and interwoven within an ecology of a socio-political industrial complex, that progressive action to end it seems impossible. It has become what King describes as a crisis that only protest can force an apathetic community to confront. The Groundswell Approach appeals to this frustration and no group in the world has answered the call to demand action on this issue like the global community of youth. Led by Greta Thunberg's Fridays for Future marches and fuelled before, during, and after each protest by social media and digital media dissemination (texts, emails, blogs, file-sharing, etc.), this chapter will explore how this generation of youth is using digital technology to unite each other, inform each other, document and report what traditional media might overlook, and, most importantly, persuade and influence those in power to address an issue that can no longer be ignored.

Media theorist Henry Jenkins believes the rapid growth of the Groundswell Approach can be traced back to a combination of political inaction on the climate change crisis and a facility for engaging with powerful digital media that the modern-day youth takes for granted:

> On the one hand, the institutions historically associated with American democracy are dysfunctional, public trust in core institutions is eroding, civic organizations no longer bring us together, elected representatives are more beholden to big contributors than to voters… and very little is likely to emerge at the level of institutional politics that is going to shift conditions very much. On the other hand, we have seen an expansion of the communicative and organizational resources available to everyday people (and grassroots organizations). (Jenkins et al. 3)

## 3.1 Social Media

One of the most powerful communications systems on the Internet is social media and as such, plays an "indispensable role" in fuelling youth's activism (Yeun and Tang, 3). Social media platforms have evolved, and continue to evolve, to best serve the needs of social movements like climate activism. This also represents one of the most significant ways the global community of youth is engaging in the Groundswell Approach to globalize its message and exert pressure on governments and industry for immediate action.

It is essential that members of any social movement have a common identity. This goes beyond the obvious goal of ending climate change, and has more to do with representation and public image to attract new members and to mobilize protests and related actions. Scholars such as Christoph Herrmann et al. believe there are as many as three identities necessary for any successful social movement:

> The *individual identity*, such as age or personal lifestyle, which brings like-minded people together and affects satisfaction in a social movement, the *collective identity*, such as the feeling of a moral or political connection to a larger group or a particular action and the *public identity* as 'activists shared definition of their situation, the expressive character of all action, the affective bonds that motivate participation, the experience of solidarity within movements and others'. Second, a social movement must evolve the ability to organise its protest and mobilise and communicate with its members. Social media is important in meeting the two requirements and enabling the organisation of actions and communications within the movement and with the public, while enhancing the identity of a social movement (Herrmann et al. 3).

These identities are key to defining the membership, goals, and impacts of youth climate activism, as well as the movement itself. These identities form a participatory culture that showcase and encourage other forms of media for expressions being the platforms of social media. Producers of such media products as documentary film (YouTube), call-to-action videos (TikTok), memes, artwork, and photography (Facebook, Twitter, and Instagram), are all forms of media that find their own lives outside of social media and further extend the reach of youth climate activists to the general public and government. The aim here is to migrate the movement's participatory culture to the arena of participatory politics "by any media necessary" (Jenkins et al. 2016). According to research conducted by sociologists Jacquelien van Stekelenburg and Bert Kandermans, these digital networks strengthen collective identity because online anonymity and reduced social cues decrease perceived differences among members, fostering the group's unity, identification, and solidarity (van Stekelenburg and Kandermans, 4). In other words, it is much easier to be inclusive and democratize youth climate movements with the relative anonymity of its members.

This wide range of media representations provide content for the social media platforms as well as provide a more inclusive population comprised not only of passionate activists who compose and deliver compelling speeches, but also other equally passionate activists who express themselves in less public ways. Just as a picture is worth a thousand words, the more introverted youth climate activists can be heard through these digital networks and related artistic productions. Social media provides a platform for causes, to be sure, but also for multimedia creations by all, denying no-one a voice on a global scale. This inclusive telecommunications system ensures maximum volume of the voice of the global community of youth and is opening the door to negotiation with those in power. However, while collective action, whether through global social media campaigns, large-scale protests, or signing petitions, may be necessary to garner the attention of politicians, it is certainly not sufficient for generating legislative action (Chia, 21).

Members of the most popular platforms can engage with each other based on selecting “friends” or becoming “followers” of those they “like” or whose posts line up with their own ways of thinking. According to a January 2022 study conducted by Statista, a carbon-neutral German company specializing in market and consumer data, the most popular social networks worldwide ranked by number of monthly users (in millions) are:

- Facebook (2910)
- YouTube (2562)
- WhatsApp (2000)
- Instagram (1478)
- Weixin/WeChat (1263)
- TikTok (1000)

Surprisingly, Twitter comes in at 15th spot in this study with a monthly user number of 436 million, falling behind Pinterest with 444 million, Snapchat at 557 million, and Facebook Messenger at 988 million, among others (Statista 2022). Since Thunberg launched her school strike through Instagram (14.4 million followers) and Twitter (5 million followers), much of the youth climate movement have congregated and interacted with each other and government officials through these digital spaces. It would appear that Instagram and Twitter are the preferred social media platforms for youth climate activists.

Twitter, in particular, is fertile ground for developing storylines and storytelling, a form of rhetorical ecology shared among youth climate activists. In a study conducted by Emily Wielk and Alecea Standlee entitled *Fighting for Their Future: An Exploratory Study of Online Community Building in the Youth Climate Change Movement*, they identify distinct narratives and patterns common in creating an online community for fellow climate activists:

> We identified three major discursive narratives that demonstrate the use of Twitter to develop community and create motivations for mobilization around global climate issues. First, our findings suggest that the strategic presentation of self by the online activists exists along a rhetorical spectrum that ranges from "full-time activist" to "teenager with an activist orientation." Secondly, activists employ a narrative that creates community in both digital and offline spaces, utilizing the specific characteristics of Twitter, such as hashtags, retweet functions, character limits, and embedded multimedia. Finally, we identified patterns of communication that focused on the deployment of emotional language and evoked emotional connections as a call to action. Taken together, these patterns underscore the degree to which technology, discursive narratives, and strategic communications create a sense of global community. (Wielk and Standlee, 27)

## 3.2 The Hashtag

Instagram, on the other hand, is a social media platform originally intended to share pictures "instantly" taken on users' cell phones, and is second only to YouTube as the most popular social media platform for teenagers in the United States used by 72% of US teens with cell phones (Jiang and Anderson 2018). Specifically, Instagram defines itself as "a free photo and video sharing app available on iPhone and Android" available for "anyone 13 or older" (Instagram Help Center). One of the most popular features of this platform is the hashtag (#). Originally suggested by Chris Messina in 2007 for Twitter use (Messina 2007), it serves as a collective keyword for comments and posts on a specific subject. In the youth climate movement, hashtags such as #fridaysforfuture and #schoolstrikeforclimate are seen as calls to action and others such as #earthday or #worldenvironmentday direct users to information, other youth climate activities, and events. The hashtag on Instagram in particular reveals another manifestation of the Groundswell Approach, this time of the hashtag itself. Using #fridaysforfuture as the principal hashtag for Instagram posts related to youth climate protest marches, one will find many more stemming from this. In a 2022 study conducted by Herrmann et al., here are the top nine of the many hashtags that emerged from the central hub of #fridaysforfuture:

- #climatestrike
- #climatechange
- #climatestrikeonline
- #gretathunberg
- #climatecrisis
- #shotoniphone
- #climateaction
- #globalwarming
- #schoolstrike4climate

Only one hashtag—#shotoniphone—is not directly related to the cause of climate activism. The diagram below reveals the vast number of hashtags in this cluster that are directly linked to the hub hashtag #fridaysforfuture. While many interconnected hashtags on Instagram are related directly to the climate movement, we can see how other more established environmental organizations like Greenpeace and Extinction Rebellion are joining forces through this method in this hub chart created by Herrmann et al. for their study *#fridaysforfuture—What Does Instagram Tell Us about a Social Movement?* (*see* Fig. 3.1).

Another study conducted by political anthropologists Yarimar Bonilla and Jonathan Rosa finds that the hashtag serves two purposes when used by youth: one is as a form of cataloguing, making certain aspects of their movements—people, events, places—easy to access with specific information related to those highlights and are easy to source. The other is more sublime. The creation of the hashtag emphasizes the importance of a particular action, individual, or even a philosophical ideal within the movement that directs its members to a sharper focus, a form of micro recruitment within the macro scope of the cause. For example, the hashtag *#fridaysforfuture*, lets youth climate activists know that this specific event is important among the many similar events that may exist or be planned throughout the movement and that a focussed support of this event is

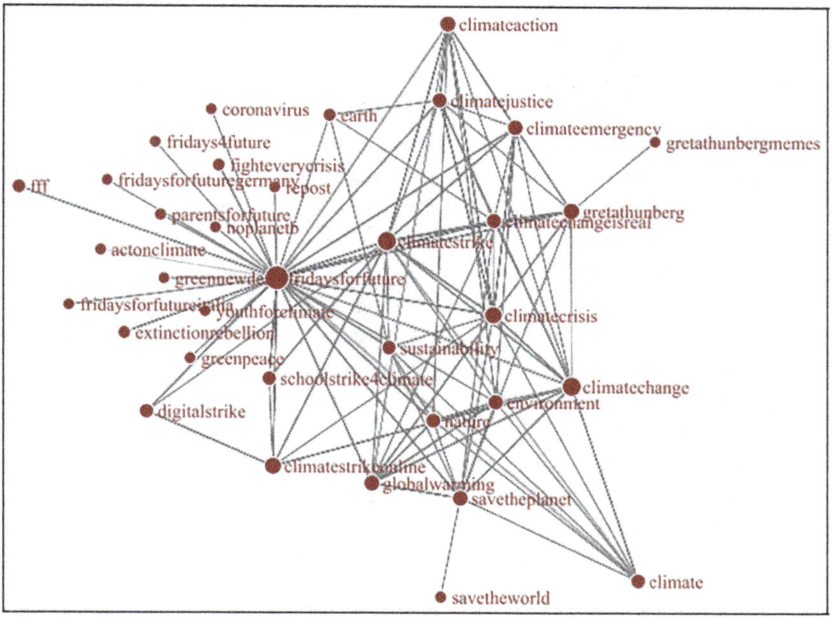


**Fig. 3.1** Instagram Hashtag Cluster (Herrmann et al. 7)

required by reading and sharing the content within the hashtag's database as well using the hashtag itself to similarly alert members of the cause in their own networks.

> The hashtag serves as an indexing system in both the clerical sense and the semiotic sense. In the clerical senses, it allows the ordering and quick retrieval of information about a specific topic…But, in addition to providing a filing system, hashtags simultaneously function semiotically by marking the intended significance of an utterance. Similar to the coding systems used by anthropologists, hashtags allow users to not simply 'file' their comments but to performatively frame what these comments are 'really about,' thereby enabling users to indicate a meaning that not be otherwise apparent. (Bonilla and Rosa, 5)

The hashtag represents another mechanism in the toolbox of the Groundswell Approach for this generation's version of activism, and while effective in attracting new members, building numbers, and focusing messages, there is some doubt about its efficacy to the goals of the movement. Some scholars see the hashtag as "something that gives you a nice feeling…(but) is completely divorced from actual legislation and governance" (Ghosh, 130–131). When used by itself without any additional text, the hashtag is seen as the ultimate example of "slacktivism" (Simpson, 240), a lazy form of protest that contributes to the movement in the least effective manner and represents members of the movement who are only active online (Morozov 2011).

Drawing a straight line between the use of hashtags by youth climate activists and actual impact among changes in government policies is rare without the policymaker specifically citing an influential hashtag that led to their decision to enact progressive legislation. One example of a presumed connection between hashtag activism and significant change arises from the Fridays For Future campaign waged in Germany in 2020. With the COVID-19 pandemic locking down most communities worldwide, a planned event for April 24, 2020 took the form of a digital strike. As such, promotion of the event was almost exclusively conducted online and the use of the hashtag #fridaysforfuture was encouraged to be used in all tweets and posts. A study conducted by the Ludwig Maximilian University of Munich found that "due to the high numbers of protesters, the strong organizational foundation, and the presumed impact on one of the biggest greenhouse gas emitters, Germany can be viewed as a cornerstone for the overall success of FFF considering the potential to induce change at all necessary levels: hearts and minds, as well as policies and laws" (Haßler et al., 5).

Another study presented at the International Conference on Information and Education Technology in 2021 looks at youth climate activists using hashtags on both Twitter and Instagram to reveal that Instagram is outpacing Twitter as the preferred platform for young climate activists: "The type of social network that predominates among teenage activists is Instagram rather than Twitter (Fernández-Prados, 482). The study also finds that most of these activists are female with as many as 74 per cent of the top 38 environmental cyberactivists between 17 and 19 years old on both platforms" (Fernández-Prados, 481).

It would appear from the research in this chapter that Instagram and Twitter are not only the preferred platforms for youth climate activists, but that predominantly female users have incorporated the affordances of these networks to engage with each other globally, index and identify as important key events and people, and to recruit new members to cause as well as to recruit existing members to act.

## 3.3 Recruitment

It may be surprising to learn that social media event invitations represent a small percentage of the recruitment process in attracting youth climate activists to attend protests and marches. In fact, there is a substantial body of scholarship that bears this out: Hautea et al. (2021), Wahlström et al. (2019), Fisher (2018), Mercea et al. (2013). As we saw in the previous chapter in the 2020 study conducted by MacEwan University (Edmonton) and Université du Québec (Trois-Rivières), recruitment to attend specific climate rallies and protests only accounts for less than 5% of posts on Twitter. A report in a recent article published in *The Conversation* supports this data:

> Rather than trying to use social media to organise events and actions, environmentalists increasingly use it to demonstrate how many other people share a cause. This allows people to take action in multiple places and times, with many encouraged to join by the number of like-minded people they see online. (*The Conversation*, 2022)

This might seem surprising since "event postings" are usually used these days to invite participants for most social activities such as parties. While research suggests that majority of participants in climate marches did so after being "personally invited" by friends, family, and teachers—as many as 70.5%, according to one study (Wahlström, 12)—most mobilization of youth on the Internet and through digital media platforms seems to start with Social Movement Organizations (SMOs) who "are known to play an important role in recruitment" (van Stekelenburg and Kandermans, 1). In a 2017 study entitled *Protesting Youth: Collective and Connective Action Participation Compared*, the authors find that one of the central appeals for digital platforms for protest recruitment—and likely the reason for its success—is that it offers the users the opportunity to be producers of content. Unlike writing a letter to the editor of a newspaper which may be edited mispresenting or decontextualizing the author's original intent or end up not being published altogether, the Internet offers production means and resources to manufacture content that aids in the influence and persuasion to attend climate protest marches.

> Social media make self-organized connective action feasible by on the one hand lowering information, communication, and coordination costs, and on the other hand offering a space and tools to produce, express, and perform

> political dissent. Compared to traditional media and the Internet, social media involve a distinctive media experience. In addition to Internet spaces where opinions and information are exchanged, social media also offer Internet tools that are used for mobilizing people and to inform and stay informed during the protest. Thus, they offer media consumers the possibility to act as *producers*. (Stekelenburg and Kandermans, 3)

SMOs and young climate activists use these recourses in producing content that becomes shareable, or as digital media theorist Henry Jenkins describes it, "spreadable" media (Jenkins 2013). More commonly understood as simply "posting" or "sharing", online content disseminated to others takes on a nature of "spreadability" with its intent of circulation and its enhancement through user production.

> We may share songs from our favorite band as a way to define ourselves, to communicate something about who we are and what we like to our friends. We may pass along a news article to a former schoolmate to strengthen our social ties with her, to remind her we remember what she is interested in. We may include a video clip in a blog post or Twitter update as a means to provide commentary on it, using that clip as inspiration for our own content and as a means to gain our own notoriety and audience. Or we may share material as a way to grow or activate a community, whether that be passing along a television-show clip for fellow fans to help dissect or spreading a protest video to mobilize or recruit others around a social cause. (Jenkins, Ford, Green, 199)

The latter example provides one of the successful methods of propaganda used by youth activists and SMOs to reinforce their followers' beliefs and passions in their cause or to recruit new membership. By introducing photography, video, and other multimedia of previous protests to an invitation to attend a new one, the visuals support the inference that "everyone is coming, so don't be left out", resulting in a more trustworthy "interpersonal" recruitment. Instead of a direct invitation from climate protest organizers, awareness of the event comes from DMs (Direct Messages), email, and social media posts from personal contacts and preapproved social media "friends" and "followers".

> Interpersonal recruitment is not just a matter of personally being asked (being the target of a recruitment attempt) but may also involve personally asking others (recruiting others). Significantly more respondents indicated

> that they had asked others themselves: 72.4% of surveyed school students personally asked someone else to join them in demonstrating at the Global Climate Strike. For adults, this share is lower, but still high at 56.3%. Relatively more demonstrators are thus recruiters themselves, rather than (solely) being a recruit. Recruiters were mostly those who did not receive a personal invitation themselves: 67.9% of all recruiters had not personally been asked by someone else. Most recruiters are thus starting an interpersonal invitation chain, rather than extending an invitation. (Wahlström, 12)

This methodology is intrinsic to the success of climate protests, marches, and rallies like those organized by the transnational offices of SMOs such as Fridays For Future or as it was originally known, Skolstrejk för Klimatet (School Strike for Climate). On March 15, 2019, the first global protest of youth organized by Thunberg's group attracted more than 1.4 million young people from 2233 cities and towns in 128 countries, with demonstrations held from Australia to India, the UK, and the USA ("School Climate Strikes", *The Guardian*, 2019). Just five months following the inaugural global event and its impressive numbers, a subsequent event held worldwide on September 20, 2019, also organized by Fridays For Future, took place attracting more than six million people in 185 countries "united across timezones and cultures" in what became the biggest climate protest in history ("Across the Globe", *The Guardian*, 2019).

As impressive and as record-setting as these numbers are, for youth climate activists like Thunberg, they are still waiting for the governmental and intergovernmental action they expected these global protests to cause. Change of this magnitude cannot be expected to take place overnight and one must recognize the significant impacts these protests did yield. The scientific community and the United Nations responded almost immediately.

> The movement strengthened and became institutionalized by gaining the support of 350 Dutch scientists who signed an open letter endorsing the school strikes in Holland. In England, in the same period, another 244 academics showed their support to the cause. The United Nations officialized its contribution to the movement by creating a specific platform for young world leaders - UN Youth Climate Summit - in order to facilitate and strengthen the movement and build space for debate, exchange of information, and recognition of the actions of activists. Some of these young people managed to speak at the Climate Summits and became UN consultants. (Monteiro 2019)

Social media platforms around the world were trending these marches and reporting on the ensuing impacts. With the power of media coverage in the hands of the youth, the events did not have to concern themselves with publicists to engage mainstream media. The coordinated efforts of organizers and protestors were successful in creating a global protest held simultaneously in almost every country in the world on an issue of global concern. The historical tendency to ignore the contributions and voices of youth (Anderson 2016) no longer applies. The media coverage generated by participants and those professional reporters who cover such events for conventional media channels ensured the voices of youth were heard.

## 3.4 Informing

The digital domain again presents itself as a new space where the global community of youth are congregating, this time to use its applications and platforms to inform themselves, each other, and those in power on the subject of climate change. Traditional methods of educational engagement are "alienating" young people as they fail to acknowledge and adapt to the creative, critical, and interactional ways in which citizenship is and might be expressed by young people who have come to source instant information in the digital age (Tupper, 87). This coincides with other studies in which students feel the school curriculum is "marginalizing" as it fails to acknowledge their differences in terms of poverty, Aboriginal cultures, immigrants and newcomers, rural/urban divides, mental health, and so forth (Tilleczek et al. 2013) and to bring them into meaningful discussions about climate change, the related societal changes required to mitigate our contributions to climate impacts, and the related global policy.

Another study examining why youth are searching for answers outside of the school system finds that youth today are seeking information on climate change from external sources because they perceive it as a direct threat to themselves. The information they seek is to further their understanding of the issue and thereby find ways that can be introduced into the society they are forming in the years to come.

> Adolescents' propensity to seek information about climate change is associated with their own risk perceptions and response efficacy. The responsive group, members of which recognize both the dangers of climate change and the potential for action to reduce the threat, is the group with the highest motivation to seek information on the issue. (Mead et al, 46)

These challenges of digital and cultural isolation as expressed by today's youth also offer the solutions being embraced by the global community of youth. They are turning to each other, not just in the classroom, but globally through the Internet to seek understanding, to inform each other, and to mobilize a united front to demand legislative and systemic changes to the way governments are currently allowing our anthropogenic influence to terra-form the planet they will soon inherit. The Mead study also looks at the role parents play in informing their adolescents and guiding their related beliefs and behaviours on climate change data and activism. The study recognizes that this period has historically been marked by "rebellious" attitudes towards parents when they intervene, but this traditional behaviour does not seem to apply when global environmental crises are being discussed.

> Adolescence is often thought to be a period characterized by rebellion against parental authority, which raises the possibility of attenuated associations between parental and adolescent attitudes and behaviors. Empirical evidence indicates, however, that this is by far the exception rather than the norm. (Mead et al, 45)

While education systems worldwide may be marginalizing youth with their approaches to engaging students of different backgrounds and curricula that do not address climate change science and related geo-political impacts due to outdated (or more precisely, non-updated) methods, approaches, and content, studies are showing that youth are adapting to these needs in their information-seeking outside of the relatively archaic classroom.

> Those with positive attitudes toward environmental issues, high-risk perceptions, and high response efficacy were more likely to seek information about the environment, regardless of their family background or demographics. From an audience segmentation perspective, this indicates that a focus on psychosocial factors, more than on demographic indicators, is likely to bear more fruit in promoting the adoption of behaviors that promote environmental sustainability. (Mead et al, 45)

Their digital library, the Internet, is a great equalizer. It democratizes all users regardless of gender, culture, race, class, and location, provided access to equipment and Wi-Fi signals are available. And since many youth are perceiving the climate crisis as a problem that will directly impact on them more than their parents or any other elder in their respective societies, they are motivated to surf the web to find information and answers that they can share with each other and use to persuade authorities who fail to share their sense of urgency.

Youth climate activist Jamie Margolin expresses her generation's frustration with the education system's approach to climate change information and activism in her book, *Youth to Power: Your Voice and How to Use It*, by stating the current school system is designed to repeat information taught in the classroom and dissuades critical analysis of it:

> From the minute we are tossed into preschool we are told 'Be quiet, raise your hand of you want to speak, listen to authority always, put your head down and do your work, and never question anything you are told. Memorize information and regurgitate it back onto a piece of paper.' We are told 'If you follow the rules, there is a clear and safe path ahead of you: you get good grades and you'll be successful. Study hard for a bright future.' But what happens when there is no future to study for? What happens when, for example, your generation is being left with a planet that is soon going to be unable to sustain human civilization because of climate change? (Magolin, xiii)

The book that Greta Thunberg describes as a "toolbox" in its foreword, serves as a framework for youth activism to end climate change, a blueprint of methods and approaches intended to answer the question "what can I do" that so many emerging environmentalists asked when confronted by a global problem as seemingly insurmountable as climate change. With chapters addressing "first steps", political writing and how to get published, heading up your own movement, the effective use of art, organizing events, "becoming the news", and building communities, *Youth to Power: Your Voice and How to Use It* is one example of many ways youth are speaking to each other in constructive, artistic, and collaborative ways to address climate change as a global community.

Traditional intergenerational relations also contribute to the disconnect between what young people perceive must be done and their elders' relative inaction. Youth often do not have the experience, credentials, or the authority to exercise equitable power dynamics with those in power resulting in little to no change in corporate and governmental behaviours and policies with respect to the environment. If things are going well, and profits are being made, industry is reluctant to change. Coupled with the intrinsic symbiotic relations between capitalism and politics, government officials are also reluctant to change as corporate interests and their related political support ensure job security for the elected officials. Unlike the rare exceptions this book has examined in the banning of DDT and CFCs, there are few examples these days of transnational governments doing the same again for an issue as globally pervasive as climate change. This inaction is at the core of the youth climate movement and the aim of its protests as an evident injustice to future generations is being committed in the face of what science has clearly demonstrated as practices that are destroying the sustainability of the planet. In her book, *Climate Change and Intergenerational Justice*, Tracey Skillington defines how this injustice against youth currently operates:

> The knowing imposition of ecological harm through excessive pollution and the refusal to modify rates of resource depletion in response to wider changes in natural resource availability is clearly an example of exerting influence over others in ways that interfere with their long-term welfare prospects. In particular, it is an injustice against future generations who are unable to challenge such behaviours or to avoid their detrimental consequences. (Skillington, 28)

Millennials realize that this kind of aversion to change, no matter how objectively necessary, must be met with new methods of engagement and information exchange. While global protests clearly raise a level of awareness around the world, other forms of the Groundswell Approach are being created to instigate more meaningful impact. One of the ways is a variation on the old adage "if you can't beat them, join them". Young people are taking advantage of the affordances of the Internet to become members of the institutional communities whose messages come with greater degrees of authority and credibility, areas often lacking in many youth climate activists because of their age. Two professions this chapter will examine in this regard is journalism and science. By joining these forces, the Groundswell Approach now extends to control the message and contributes to its delivery.

## 3.5 The Citizen Journalist

In their "think paper" for The Media Center at The American Press Institute, *We Media: How Audiences are Shaping the Future of News and Information*, Shayne Bowman and Chris Willis define citizen (or participatory) journalism as "the act of a citizen, or group of citizens, playing an active role in the process of collecting, reporting, analyzing and disseminating news and information. The intent of this participation is to provide independent, reliable, accurate, wide-ranging and relevant information that a democracy requires" (Bowman & Willis, 9). As people around the world began writing their own weblogs (more commonly known by the abbreviated "blogs"), it was clear the intent was not necessarily a goal. While many bloggers are certainly independent, frequently writing on their own websites, and clearly not accountable to any editor or publishing authority, the content they produce is not necessarily reliable, accurate, wide-ranging, or relevant. Personal blogs are commonly about personal opinions on the subjects of fashion, food, travel, and current events and with the exception of the rules of engagement for many social media platforms and specific crimes addressing hate and abuse, anything goes on the blogger's own website in an ultimate expression of the right to freedom of speech.

According to WordPress, a public website creation platform, the top ten subjects of its 60 million blogs are:

1. Fashion
2. Food
3. Travel
4. Music
5. Lifestyle
6. Fitness
7. Do-It-Yourself (DIY) Projects
8. Sports
9. Finance
10. Politics[1]

With the possible exception of Politics (#10) and News (#16), blogs with environmental themes do not appear to be a universally popular subject for bloggers. So where do young climate activists blog about climate change? Mostly, on SMO and NGO sites such as Youth4Nature, Climate Generation, Climate Reality Project, Voices of Youth, Youth Climate Lab,

[1] Data provided by wpbeginners.com, March 31, 2022. Accessed: April 17, 2022. Link: https://www.wpbeginner.com/beginners-guide/which-are-the-most-popular-types-of-blogs

and even some UN-supported agency sites such as YOUNGO, the official youth constituency of the United Nations, and the United Nations Framework Convention on Climate Change's Youth Climate Report (YCR). Many of these groups support participatory journalism through membership postings. What makes a self-authored blog participatory is the interactive comment feature afforded by many of these online blogging platforms. Open, public discussions are invited by readers and edited by authors, much the same way letters to the editor are invited in print journalism and edited by magazine and newspaper editors.

According to *optinmonster.com*, a blogging research and resource centre, as of 2019, there are more than 600 million[2] blogs on the Internet—that's a lot of citizen journalists. This figure represents an increase of 12% since 2015. The average blog post is 1151 words, an increase of 42% in the last five years.[3] Since we have established that the Internet and its connected digital devices represent the preferred choice of communications for Millennials, it should come as no surprise to learn that more than half—53.3%—of the total blogging population is between the ages of 15 and 35, followed by those 20 and under at 20.2% (Levine 2010). Nearly three-quarters of all 600 million bloggers on the planet can be defined as youth. These numbers reflect a different dimension of the Groundswell Approach: interactive data collection and dissemination by the global community youth acting as citizen journalists. They aim at informing each other while at the same time providing valuable youth perspectives and data to those in power by covering aspects of the climate change story conventional media may overlook.

However, one of the perceived problems in this manifestation of the Groundswell Approach is how it is sourced by the citizen journalist and their cause in the polarizing world of political partisanship. Events such as the Fridays for Future marches gather their numbers largely through their own infrastructure, much like Black Lives Matter events, but when there is a cross-over in citizen journalist reports by youth climate activists in social media, and in particular, through related organizational structures and their public spaces, their reports attract engagement from radical opponents, belonging to a sociological category known as *countermovements* (McCarthy and Zald, 8). And while this may appear

[2] Data provided by 99Firms.com, n.d.

[3] Data provided by optinmonster.com, January 7, 2022. Accessed: April 17, 2022. Link: https://optinmonster.com/blogging-statistics

counter-productive to the cause they are fighting for, it actually grows the movement in global awareness making "both sides" participatory members. In a study conducted by Elizabeth Eide and Risto Kunelius entitled *Voices of a Generation the Communicative Power of Youth Activism*, while acknowledging the possible negative impact on the policymaker, they suggest the engagement may actually yield positive results in the long run.

> The polarization of climate debates has been a key part of the global political opportunity structure that has shaped the movement. Aggressive reactions from one side of the climate debate may have built public acceptance and sympathy for the movement. At the same time, such alliances have created challenges to the whole political elite. (Eide, Kunelius, 6)

As aggressive as the rhetoric may seem to be at times, it is a far cry from the police riots that characterized the youth protests of the 1960s. Global interconnectivity through the Internet has yielded greater numbers of followers, participants, and activists in the fight to end climate change, but when they gather and "the whole world is watching" like never before, the protests are relatively peaceful. The ubiquitous presence of cameras in the hands of protestors is a defining feature of the marches and the approach taken to citizen journalism of this generation of youth. Any bad behaviour by either side in public spaces is instantly documented and broadcast worldwide. Much like a television news network's *breaking news*, *as-it-happens* reports, and *going live*, the mobile recording and broadcasting devices prevalent in every young person's cell phones makes everyone more accountable for their words and actions greatly reducing the instances of violence that characterized youth movements prior to the Internet.

One of the world's leading mobilizers of the citizen journalist for the youth climate movement is a program called *Young Reporters for the Environment* (YRE), operated out of the Foundation for Environmental Education (FEE) at the University of Copenhagen. Established in 1981, the FEE is the world's largest environmental education organization, with more than 100 member organizations in 81 countries. Online, they opted for the extension "global" rather than the more commonly used dot-com or even dot-org to emphasize the organization's scope and dedication to inclusivity. YRE and its other school-based programs—EcoSchools and Learning About Forests (LEAF)—engage over 20 million young people in environmental education and positive action every year (FEE, 2022). While not fitting the mould of a traditional social movement organization, the 40-year-old organization uses education to arm the youth climate activists with professional approaches to citizen journalism. They do not organize marches or target specific political protests or actions, but rather provide professional instruction on journalistic research techniques and programs that allow them to apply these skills in the production of photojournalism studies, reporting, and broadcast journalism. Each year, YRE holds a competition among its citizen journalists in various categories:

- *Article*—maximum three entries (one for each age group: 11–14 years, 15–18 years, 19–25 years).
- *Single Reportage Photo*—one entry, regardless of the age.
- *Single Environmental Campaign Photo*—one entry, regardless of the age.
- *Photo Story* of 3–5 Photos—one entry, regardless of the age.
- *Video*—three entries (one for each age group: 11–14 years, 15–18 years, 19–25 years), regardless of the type of video.
- *International Collaboration*—all entries are welcome—simply make sure that only one country uploads the submission (*Competition* 2022).

The program boasts more than 450,000 young reporters in more than 40 countries worldwide. Its annual competition attracts an average of 19,000 entries each year (*A Brief History*, 2022). This version of the Groundswell Approach recruits global participation through educational programs designed to provide the youth climate activist the tools and skills their critics often say they are too young to possess. The large numbers reported by the YRE program not only indicate a strong organizational structure, but also a strong desire by young climate activists to be better prepared to walk through the doors of power and counter the criticism their generation usually gets from those in power who do not take them seriously. In addition to assisting youth with journalism skills, the YRE program also helps facilitate access to the United Nations and its policymakers through two key partnerships: the UNESCO *Green Citizens* program and the United Nations Framework Convention on Climate Change's (UNFCCC) *Youth Climate Report*. These relationships are essential portals to meaningful climate change contributions made by youth to global climate change policy. It is through portals such as Young Reporters for the Environment that the Groundswell Approach funnels into the Direct Approach for many youth climate activists.

The second half of this book will address the Direct Approach in much more detail, but for now, it is a method that emphasizes a collaborative relationship between the youth activist and the policymaker. It also varies from the Groundswell Approach by having the global community of youth talk *with* the policymaker instead of *to* them. Vocal protests around the world attended by millions of activists are effective at raising awareness of climate change issues and this generation's desire to resolve them, but that only goes so far in manifesting the impact intended by such protests. As Thunberg so memorably quipped in her September 28, 2021 address at the Youth4Climate event in Milan, in the absence of climate action by those in power, anything else is just "blah, blah, blah".[4] Measurable results are more common, however—especially with the United Nations, but also with national and regional governments—within the framework to be described and examined in Chap. 5 that this book defines as the Direct Approach.

In looking at the United Nations as a multinational governing body designed to introduce global policy on global issues, youth climate activists need to establish a relationship with this organization. The Young Reporters for the Environment program has done that through its partnership with UNESCO's Green Citizens program and the UNFCCC's Youth Climate Report. The Green Citizens program showcases climate initiatives around the world that they support, effectively endorsing these programs at the UN level for others around the world to join or to emulate: "This platform offers opportunities for partnerships and shares best practices, innovations and ideas. It is a spotlight on human adventures, which show the same fervour and shared vision of the future of our planet" (About 2022). The program currently showcases 109 projects from 56 countries and one of them is the Young Reporters for the Environment program. The UNESCO website describes it as "a platform (that) allows (youth) to research environmental issues and promote solutions through investigative reporting, photography, and video journalism" (Projects 2022).

With a full description of the project, samples of the work done by participants, and links to the YRE website and contact email, visitors to the UNESCO site, including UN policymakers, become familiar with the

[4] Full speech: https://www.nationalobserver.com/2021/09/28/news/greta-thunberg-bashes-blah-blahblah-cop26-speech

program and its products. For those who may not have known about this citizen journalism program and who prefer to seek such programs only through UN websites, this partnership and subsequent showcase provides the credibility for the program and its reports that those in power prefer to access.

The UNFCCC offers its endorsement in a similar manner to another citizen journalism project known as the Youth Climate Report (YCR). This curation project serves as a database of short documentary films made by the global community of youth aged 18–30. The project displays more than 600 video reports on climate research, impacts, and solutions on all seven continents by way of an interactive Geographic Information System map of the world. The project curates its content through a series of training sessions, workshops, and competitions.[5]

The connection between youth and UN policy is through the medium of film. As video reports are filed with the YRE competition, they are subsequently reviewed and added to the YCR project. These films are then presented directly to policymakers, heads of states, and other delegates of the annual UN climate summits known as the COP conferences each year. They are presented at various side events, special screenings, and press conferences providing a direct connection to the policymaker inside the Blue Zone of the COP conference and ensuring "more youth participation in decision-making process", a demand frequently identified by youth climate activists (Bannon 2021; *Global Youth Statement*, 13).

The role of the citizen journalist is to provide specific reporting on issues identified by the policymaker. With direct collaboration between youth climate reporters and UN policymakers, a symbiotic relationship develops through this Direct Approach: the youth climate activist engages in meaningful dialogue with those in power, and the UN policymaker is informed by the global community of youth on climate-related matters that may not be readily available through the traditional channels of scientific reports and conventional media. In addition to this relatively new role being played by young climate activists is a similar persona that addresses another criticism of youth that claims they do not have the educational credentials to comment on climate science. In this new role is the citizen scientist.

[5] http://youthclimatereport.org/film-workshops

## 3.6 The Citizen Scientist

Citizen Science (CS) is basically understood as scientific research conducted by non-scientific citizens and their contributions—often in the form of data collection in the field—are often referred to as "public participation" in the scientific process. Recent scholarship indicates that in the last 20 years there has been a significant rise in the number of scientific reports that include citizen science as a participatory research partner in formal scientific papers (Jordan et al., 2015; Shirk et al., 2012).

The recent rise in citizen science research and data collection is largely due to the importance of the contribution. With a global environmental crisis as big as climate change, scientists can only cover so much ground. For scientists studying the impacts of climate change among birds, for instance, the scientists of *eBird*, a crowd-sourced database using the global community of bird-watchers as citizen scientists, took the community approach involving the active participation of all stakeholders, including the citizen scientist whose field work is crucial to the database of a project designed to account for all the birds of the world, their locations, and migratory patterns. This participatory perspective revealed a new methodology of data collection and management to the scientist (Terry and Hewson 2022): "Ultimately, these collaborations have enabled us to increase both the quantity and the quality of useful data available for analysis. Further, this broad spectrum of intellectual contributions and applications has fundamentally changed our view of the project, which we now see as a collective enterprise" (Sullivan et al. 2014, 31). With this additional data, more accurate findings can be made that better service the policymaker. In effect, the citizen scientist bridges the gap between society and academic research (Gunk et al., 149) and this, in turn, democratizes the process of scientific study and policy creation.

Along with these benefits come problems inherent with using unskilled labour in the performance of scientific research. Will the reported data be accurate? Are some of the observations tainted by being made with untrained eyes? What are the ethical considerations of using non-scientific participants in scientific work? To address most of these concerns, the citizen scientist does not usually work without being in direct collaboration with the scientific community they are serving (Cheung et al. 2022). CS by youth is rapidly expanding (Ballard et al. 2022) and a recent study found that when youth work alongside scientists, they not only perform the labour reliably under professional guidance and supervision, but they also develop a concern for the environment where none might have previously existed. They were also found to want to continue this work long after the project and their contributions to it have ended:

> We identify three key (citizen science) processes through which many of the youth developed environmental science agency: ensuring rigorous data collection, disseminating scientific findings to authentic external audiences, and investigating complex social-ecological systems. Our findings suggest that when (citizen science) programs for youth support these processes, they can foster youth participation in current conservation actions, and build their capacity for future conservation actions. (Ballard et al. 2022)

## 3.7 Conclusion

This participatory approach to scientific research and science understanding is similar to the participatory approach of collaborating with the policymaker as a citizen journalist: where there is a perceived lack of experience in youth that threatens their credibility and diminishes the import of their voice, they are not taken seriously. However, where their age fails them in experience, their participation in data collection and information gathering together with experts in their respective fields (climate science and environmental policy) gives them the confidence to speak with relative authority. This approach also serves to give more credibility to the policymaker when speaking truth to power.

## Works Cited

"About Green Citizens", *UNESCO.org*. Accessed: April 23, 2022. Link: https://www.unescogreencitizens.org/unesco-green-citizens

"About Us", *earthday.org*. Accessed: April 19, 2022. Link: https://www.earthday.org/about-us

Eide, Elisabeth and Kunelius, Risto. "Voices of a Generation the Communicative Power of Youth Activism", *Climatic Change* 169: 6, November 2, 2021.

"Across the globe, millions join biggest climate protest ever", *The Guardian*, September 20, 2019. Accessed: April 17, 2022. Link: https://www.theguardian.com/environment/2019/sep/21/across-the-globe-millions-join-biggest-climate-protest-ever

"Competition Structure", *Young Reporters for the Environment*. Accessed: April 23, 2022. Link: https://www.yre.global/competition-structure

"Foundation for Environmental Education", *fee.global*. University of Copenhagen. Accessed: April 17, 2022. Link: https://www.fee.global

"Global Youth Statement", *United Nations Climate Change/YOUNGO*, Glasgow: 2021. Accessed: April 23, 2022. Link: https://ukcoy16.org/wp-content/uploads/2021/10/Global-Youth-Statement.pdf

"Most popular social networks worldwide as of January 2022, ranked by number of monthly active users", *Statista.com*, March 8, 2022. Accessed: April 19, 2022.Link:https://www.statista.com/statistics/272014/global-social-networks-ranked-by-number-of-users

"Projects", *UNESCO.org*. Accessed: April 23, 2022. Link: https://www.unesco-greencitizens.org/projects/young-reporters-for-the-environment

"Social Media Can Support Environmental Movements – But Not in the Way You Might Think", *The Conversation*, March 8, 2022. Accessed: April 17, 2022. Link: https://theconversation.com/social-media-can-support-environmental-movements-but-not-in-the-way-you-might-think-177847

"School climate strikes: 1.4 million people took part, say campaigners", *The Guardian*, April 12, 2019. Accessed: April 17, 2022. Link: https://www.theguardian.com/environment/2019/mar/19/school-climate-strikes-more-than-1-million-took-part-say-campaigners-greta-thunberg

*Society*, Volume 7, Issue 2, April 20. 2021. Accessed: April 17, 2022. Link: https://journals.sagepub.com/doi/pdf/10.1177/20563051211012344

"What is Instagram?", *Instagram Help Centre*, Instagram.com. Accessed: April 21, 2022. Link: https://help.instagram.com/424737657584573

Anderson, Priscilla. *The Politics of Childhood Real and Imagined, Volume 2.* Oxon: Routledge, 2016.

Bannon, Laura. "COP26 – Will It Make a Difference?", *Inconvenient Questions International.* December 4, 2021. Accessed: April 23 6, 2022. Link: https://youtu.be/U8qAVjwa1h8

Bonilla, Yarimar, and Jonathan Rosa. "# Ferguson: Digital Protest, Hashtag Ethnography, and the Racial Politics of Social Media in the United States," *American Ethnologist* 42.1, 2015: 4–17.

Bowman, Shayne, and Willis, Chris. *We Media: How Audiences are Shaping the Future of News and Information.* Reston, VA: The Media Center at the American Press Institute, 2003.

Cheung, S. Y., Leung, Y. F., & Larson, L. R. (2022). "Citizen Science as a Tool for Enhancing Recreation Research in Protected Areas: Applications and Opportunities", *Journal of Environmental Management*, 305. Accessed: April 24, 2022. Link: https://www.sciencedirect.com/science/article/pii/S0301479721024154?casa_token=aRdeIBkVSoQAAAAA:9au69N93CSqic_uTcy9S-T4UsbZ4szinWVsSqItLp23hHJ2J8i67T2cObJ3GmIDEA65lUHinC8g

Chia, Justin. "Social Media and the Global Climate Strike: A Tool for Youth Climate Change Activists and Politicians," *Sojourners Undergraduate Journal of Sociology*, 12.1, 2021: 18–39.

Erin Mead, Connie Roser-Renouf, Rajiv N. Rimal, June A. Flora, Edward W. Maibach & Anthony Leiserowitz. "Information Seeking About Global Climate Change Among Adolescents: The Role of Risk Perceptions, Efficacy Beliefs, and Parental Influences", *Atlantic Journal of Communication*, 20:1, 2012: 31–52.

Fernández-Prados, J. S., Lozano-Díaz, A., Cuenca-Piqueras, C., & González-Moreno, M. J. (2021, March). "Analysis of Teenage Cyberactivists on Twitter and Instagram around the World". In *2021 9th International Conference on Information and Education Technology (ICIET)* 2021: 476–479.

Fisher, Dana R. "Climate of resistance." *The Resistance: The Dawn of the Anti-Trump Opposition Movement*. New York: Oxford University Press, 2018.

Ghosh, A. *The Great Derangement: Climate Change and the Unthinkable*. Chicago and London: University of Chicago Press, 2017.

Gunko, R., Rapeli, L., Scheinin, M., Vuorisalo, T., & Karell, P. "How Accurate is Citizen Science? Evaluating Public Assessments of Coastal Water Quality", *Environmental Policy and Governance*, 2022.

Haßler, Jörg, Anna-Katharina Wurst, Marc Jungblut, and Katharina Schlosser. "Influence of the Pandemic Lockdown on Fridays for Future's Hashtag Activism." *New Media & Society* (2021): 14614448211026575.

Hautea, Samantha, Parks, Perry, Takahashi, Bruno, and Zeng, Jing. "Showing They Care (Or Don't): Affective Publics and Ambivalent Climate Activism on TikTok", *Social Media + Society,* April-June, 2021.

Heidi L. Ballard, Colin G.H. Dixon, Emily M. Harris, "Youth-focused Citizen Science: Examining the Role of Environmental Science Learning and Agency for Conservation", *Biological Conservation*, Volume 208, 2017: 65–75.

Herrmann, Christoph, Sebastian Rhein, and Isabelle Dorsch. "# fridaysforfuture–What does Instagram Tell Us about a Social Movement?" *Journal of Information Science* (2022): 01655515211063620.

Jenkins, Henry, Sam Ford, and Joshua Green. "Spreadable Media", *Spreadable Media*. New York University Press, 2013.

Jenkins, Henry, Sangita Shresthova, Liana Gamber-Thompson, Neta Kligler-Vilenchik, and Arely Zimmerman. "By any media necessary." In *By Any Media Necessary*. New York University Press, 2016.

Jiang, Jingjing and Anderson, Monica. "Teens, Social Media and Technology 2018", *Pew Research Center*, pewresearch.org. Accessed: April 21, 2022. Link: https://www.pewresearch.org/internet/2018/05/31/teens-social-media-technology-2018/#:~:text=YouTube%2C%20Instagram%20and%20Snapchat%20are,they%20are%20online%20'almost%20constantly'

Jordan, R.; Crall, A.; Gray, S.; Phillips, T.; Mellor, D. "Citizen Science as a Distinct Field of Inquiry", *BioScience* 2015: 65, 208–211.

Levine, Sheldon. "Blogger Demographics: Who is Responsible for the Majority of Blog Content?", *Social Media Today*, June 7, 2010. Accessed: April 17, 2022. Link: https://www.socialmediatoday.com/content/blogger-demographics-who-responsible-majority-blog-content#:~:text=Not%20surprising%2C%20bloggers%20in%20the,the%20most%20bloggers%20with%2029.2%25

King, Martin L. *Letter from the Birmingham Jail.* San Francisco: Harper San Francisco, 1994.

Margolin, Jamie. *Youth To Power: Your Voice and How to Use It*, New York: Hachette Book Group, 2020.

McCarthy, J.D., & Zald, M.N. (1977). "Resource Mobilization and Social Movements: A Partial Theory", *The American Journal of Sociology*, 82, 1212–1241. https://doi.org/10.1086/226464.

McLuhan, M. (1962). *The Gutenberg Galaxy: The Making of Typographic Man.* Toronto: University of Toronto Press, Jan 1, 1962. Print.

Mercea, Dan, Paul G. Nixon, and Andreas Funk. "13 Unaffiliated socialization and social media recruitment." *Politics and the Internet in Comparative Context: Views from the Cloud.* New York: Routledge, 2013.

Messina C. How do you feel about using # (pound) for groups. As in #barcamp [msg]? 2007. Accessed April 23, 2022. Link: https://twitter.com/chrismessina

Monteiro, Laura, and Mauro Guilherme Maidana Capelari. "Contextualizing Youth Climate Activism in Brazil: Knowledge, Motivations, and Policy Actions", *International Conference on Public Policy (ICPP 5)*, Barcelona, 2021.

Morozov, Evgeny. *The Net Delusion.* London: Allen Lane/Penguin Books, 2011.

Shirk, J.L.; Ballard, H.L.; Wilderman, C.C.; Phillips, T.; Wiggins, A.; Jordan, R.; McCallie, E.; Minarchek, M.; Lewenstein, B.; Krasny, M.E.; et al. Public Participation in Scientific Research: A Framework for Deliberate Design, *Ecology and Society.* 2012, 17, 29.

Simpson, Ellen. "Integrated & Alone: The Use of Hashtags in Twitter Social Activism." *Companion of the 2018 ACM Conference on Computer Supported Cooperative Work and Social Computing.* 2018.

Skillington, Tracey. *Climate Change and Intergenerational Justice.* New York: Routledge, 2019.

Stekelenburg, Jacquelien & Klandermans, Bert. "Protesting Youth: Collective and Connective Action Participation Compared". *Zeitschrift für Psychologie*, 2017. 336–346.

Sullivan, Brian L., Christopher L. Wood, Marshall J. Iliff, Rick E. Bonney, Daniel Fink, and Steve Kelling. "The eBird Enterprise: An Integrated Approach to Development and Application of Citizen Science". *Biological Conservation,* 2013: 169, 31–40. https://doi.org/10.1016/j.biocon.2013.11.003, 2014.

Terry, Mark and Hewson Michael. *The Emerging Role of Geomedia in the Environmental Humanities.* Lanham, MD: Lexington Books, 2022.

Tilleczek, Kate, and H. Bruce Ferguson, eds. *Youth, Education, and Marginality: Local and Global Expressions.* Wilfrid Laurier University Press, 2013.

Tupper, Jennifer. "Social Media and the Idle No More Movement: Citizenship, Activism and Dissent in Canada." Journal of Social Science Education 13, no. 4 (2014): 87–94.

Wahlström, Mattias, et al. *Protest for a future: Composition, mobilization and motives of the participants in Fridays For Future climate protests on 15 March, 2019 in 13 European cities,* 2019.

Wielk, Emily, and Alecea Standlee. "Fighting for Their Future: An Exploratory Study of Online Community Building in the Youth Climate Change Movement." *Qualitative Sociology Review* 17.2 (2021): 22–37.

Yuen, Samson, and Gary Tang. "Instagram and social capital: youth activism in a networked movement." *Social Movement Studies* (2021): 1–22.

CHAPTER 4

# A Seat at the Table

In addition to protests around the world and online activism that have seen the global community of youth engage in unprecedented numbers, young people are hearing the criticisms of their voice and are taking measures to overcome them. They lack credibility, critics say (Spajic et al. 2019, 373), because they are not (yet) accredited and professional scientists; they have no experience in politics, governances, and policymaking; they are not business professionals with necessary understandings of corporate structures, goals, and macroeconomics. It is generally considered difficult to measure direct impact between protest marches and new climate policy; however, "while one cannot measure the direct causal impact that these climate strikes have had on state and intergovernmental climate change policies, this worldwide youth mobilization has aroused a sense of urgency, provided an alternative discourse, and cultivated youth leadership and commitment to civic action" (Han and Wuk Ahn 2020, 2). As well, when activated, youth, in partnership with adults, have changed the narrative around their own ability (Teixeira et al. 2021, 142), a direct connection can be made.

The previous chapter introduced the "citizen" prefix to youth activities intended to counter these criticisms, serving as interns performing the tasks in arenas they are accused of having no experience. To ensure responsible engagement, experts in these fields collaborate with these young

The original version of the chapter has been revised. A correction to this chapter can be found at https://doi.org/10.1007/978-3-031-14298-7_8


M. Terry, *Speaking Youth to Power*, Palgrave Studies in Media and Environmental Communication,
https://doi.org/10.1007/978-3-031-14298-7_4

climate activists to provide training and supervision. This, in turn, gives authority to their voice and accelerates the path to progressive and intergenerationally inclusive environmental policy change. There is value to including representation from large sectors of society who have a vested interest in policies that affect the environment in which we all live. Even without this specific mentor-led training, youth are poised to bring a unique set of skills and perspectives to the table:

> …society needs young people's open mindedness, their willingness to take risks, and their ability to innovate. Given their minimal vested interest in existing power structures, youth can question the status quo and speak truth to authority figures if given a place at the decision-making table (Spajic et al. 2019, 373).

Many of the adults in positions of power related to climate change are seen to be dragging their feet, but not because they deny climate science. The political challenges of their positions can often put up roadblocks that can seem insurmountable, especially without the necessary data and research available to the citizen journalist and citizen scientist, and the relative freedom to take risks and to innovate existing systems that current corporate and institutional structures do not allow. However, progress on this front can be made when youth are actively engaged in the policymaking process. In a 2017 report published by the Organisation for Economic Co-operation and Development (OECD), the benefits of youth policy participation impact on both youth and government: "Through participation, young people learn to process information and build decision-making abilities. It also allows young people to understand better how government bodies work and increases public transparency and thus accountability. An improved understanding of the political and administrative machinery helps to create trust in public authorities … policymakers can improve programme design and implementation by incorporating information provided by young people. Furthermore, participation increases the ownership of policies and initiatives, which is an important factor for their success" (OECD 2017). The report further identifies the problems often facing such productive engagement, citing that the main reasons for a lack of youth involvement are social, economic, and institutional in nature. The report suggests that the prevailing societal attitude towards young people is often that they are "troubled and troubling", which gives justification to create policy without their consultation or agreement.

Youth agendas set by adults who hold this opinion might focus on young people's shortcomings and problems rather than invest in youth's potential as a positive source of change (OECD 2017).

How is this prejudice reflected in government representation of youth in policy discussions? One way to examine this is through the Paris Climate Agreement's Nationally Determined Contributions (NDC). These are key components of the Paris Agreement in that every signatory nation must outline their mitigation and adaptation plans through an NDC, which is revised every five years and is expected to raise the bar to higher levels each term. National Adaptation Plans (NAPs) are also available through the United Nations Framework Convention on Climate Change to support adaptation planning for countries, with a particular emphasis on developing countries. A review of 160 NDCs and 13 NAPs conducted in 2019 revealed that "globally, countries have a long way to go in terms of including youth … in their climate strategies. This means that the rights of youth, including their right to meaningfully participate in climate policymaking that affects them, is being broadly overlooked" (Benkenstein et al. 2020, 42–43).

This chapter will look at emerging strategies of collaboration among the policymaker and youth. Specifically, an examination will be made of the times when youth are literally given a seat at the table of meaningful negotiations and discussions related to climate change policy, not the token invitation to address the United Nations, which many youth now feel is disingenuous (Spajic et al. 2019, 373). It is not as meaningful to be offered the floor to speak about your grievances as it is to sit with the policymaker and provide necessary data and community perspectives from the three key stakeholders in environmental policy creation: science, industry, government. This form of engagement is more collaborative than it is argumentative. Demanding change in heated tones is less effective than working alongside the policymaker to achieve a mutual goal. Some youth climate activists are putting down the megaphone and picking up their laptops in search of opportunities to work productively with the policymaker.

A 2019 commentary published in *The Lancet* proposes a "set of practice points" for successful youth engagement with policymakers—*consultation, facilitation, accountability,* and *evidence*—which address the key points in the framework for youth engagement proposed herein (Spajic et al. 2019, 374).

## 4.1 Consultation

"Structured, systematic consultation with young people must guide work in (climate change policy creation). Leaders in (climate policy) need to establish clear, consistent, and safe opportunities for young people in agenda setting and decision making and provide support and resources for translating youth input into tangible outcomes".

Key to this description of successful consultation are the terms "agenda setting" and "decision making" as they directly involve youth in the process of policy creation and the ensuing systemic changes. They are not seen as mere contributors to the conversation, but active participants in the process. A good example of this form of open access structure is the Prime Minister's Youth Council in Canada. The government program provides opportunities to youth across the country to serve as advisors to high-ranking government officials such as engages with the Prime Minister, his Cabinet and senior government officials several times a year in a formal capacity identified as "cohort". According to its government website, the Youth Council "provides a unique platform for youth to influence policy by expressing themselves in their own voice. Council members develop and deliver input in their own words in the medium of their choice, be it a video, a post on social media, a formal presentation, a written letter, or by using tools like e-petitions to Parliament" (Youth Council 2022). As guiding principles of participation, the government asks its youth cohorts to ensure their contributions are sourced from three main areas:

- their school, work, and life experience
- input from other youth in their community
- their research and volunteer experience (Youth Council 2022).

These criteria reflect the Canadian government's way of ensuring the voice youth bring to the table has merit being based in their own direct experiences, representing participatory consensus within their own communities, and through their own citizen reporting and research. They also address the common criticism of the lack of authoritative experience many youth have due to their age by accepting these alternatives.

Another country making progressive moves to consult with youth in its climate change policies is Kenya. Part of this country's National Climate Change Framework Policy specifically mandates that youth be part of the "decision-making" process for climate change governance:

> The youth represent a crossover between the present and future generations, and therefore play a critical role in socio-economic development, including addressing climate change. It is necessary to carve out specific roles and opportunities for youth participation in decision making in climate change governance, and to pursue opportunities that arise through climate change actions (Kenya 2017, 30).

In the Seychelles, youth are included in several climate change and environmental policymaking processes and on a variety of committees. "A consultative approach is used where policymakers engage with existing networks in the public, private and civil society sectors, including young people, to collect written submissions as part of an inter-sectoral engagement in the policymaking process… The Seychelles' National Climate Change Committee is also required to have a seat exclusively reserved for one youth member. In addition, the Seychelles National Youth Policy Forum is funded and coordinated by government with the primary role of strengthening youth priorities in policy development and implementation" (Benkenstein et al. 2020, 49).

## 4.2 Facilitation

This component of the framework recognizes that obstacles often hinder youth from participating, even when the opportunity to participate is made available. As addressed in Canada's Prime Minister's Youth Council, the obstacles of lack of experience, understanding, and professional and academic credentials need to be overcome with an alternative. Other obstacles may be financial. Some of the best youth climate activists may live in remote communities with limited to no access to the Internet and telecommunication systems like WiFi or even phone lines. Travel to places where there is access may be an obstacle if the distance is too great to travel by available transportation. Attending conferences or policy sessions in capital cities may require specific funding to facilitate their travel and attendance.

In this regard, *The Lancet* report recommends that governments and other leaders in formal climate policy positions "must facilitate youth participation by identifying and removing barriers. Many of these barriers are inherent to the adult-centric nature of institutions and decision making processes, and without necessary orientation and support young people can be left disempowered. The accessibility of opportunities across diverse groups of young people should be a key priority. Funding is another notable barrier to equitable youth participation. (Climate change) organisations must, therefore, allocate funding for youth engagement, including adequately remunerating young people for the time and expertise they contribute" (Spajic et al. 2019, 374).

The Special Report on *Youth Climate Advocacy* published by the South African Institute of International Affairs in 2020 echoes this sentiment acknowledging that while opportunities for policy participation may exist, youth "face significant barriers to (this) participation" (Benkenstein et al. 2020, 50). The obstacles identified in this report include conflicts with school schedules for youth and the prohibitive travel costs and distances for many youth. Some critics believe that a failure to facilitate meaningful participation is intentional in order to restrict the power youth may yield should they be given it.

> Youth participation in its current policy form excludes democratic participation and is restricted to a narrow definition of participation as involvement in community, cultural, voluntary and educational activities…the concept of youth participation excludes political participation and there is no commitment to increasing young people's engagement in the public sphere. Youth participation is strictly confined to volunteering, mentoring, leadership education and enterprise schemes which are directed towards the increased governance of young people (Bessant, 91).

A study out of the University of Michigan and Harvard University in the United States addresses these accessibility concerns and proposes a commission status for youth that recognizes they are "inseparable" from their communities.

> The youth commission is a distinct group whose efforts emerge in a youth-friendly environment which contributes to the scope and quality of participation… Whereas most municipal policy are dominated by adults, youth commissioners review policies proposed by public officials, set priorities through town hall meetings, and advocate their interests through face-to-face meetings with public officials. Through actions like these, they learn how to organize for political action in an arena dominated by adults (Checkoway et al. 2005, 1159).

The value in this structure is the "face-to-face" meetings with public officials. These encounters afford direct contributions to policy while at the same time providing opportunities for "on-the-job" training, as it were, for the inexperienced youth. This, therefore, represents an expedient way of addressing the criticism that youth cannot be taken seriously in policy discussion since they have no experience.

The other area often cited as a lack of experience among the global community of youth climate activists is an accredited and recognized education in climate science. We have seen how citizen journalism and citizen science address this by providing opportunities for youth to engage directly with climate science researchers, but there are more significant structures in place that provide formal educational opportunities in climate science. Research institutes and climate science organizations such as ArcticNet and the British Antarctica Survey are two examples of how university-based science groups make internships available to young people to work alongside scientists in the field providing the "experiential education" platform that is so highly valued in education today (Henderson 2004; Erickson 2013; Brooks et al. 2022). By working in the field alongside these professional and highly skilled scientists—often educators as well—the student not only learns the methodologies and practices of climate science, but is present to observe and document actual findings and results as they emerge. Being in these remote parts of the world also provides the young climate activist with an authoritative perspective of how climate change is impacting these regions.

## 4.3 Accountability

A lot has been said about holding those in power accountable for their actions—or in the case of climate policy, inaction—but little has been said about the accountability of youth. In this regard, youth accountability is measured by the specific impactful contributions they have made in consultations and engagement with the policymaker. When such engagement leads to actual policy change, no matter how small, youth feel empowered. They are encouraged by this achievement and become advocates of the process to other youth. A critical research methodology in this field is Youth-Led Participatory Action Research (YPAR), first introduced in the 2008 book, *Revolutionizing Education: Youth Participatory Action Research in Motion*. YPAR is defined as "a way to address both the political challenges and inherent power imbalances of conducting research with young people" (Cammarota and Fine 2008, 1). The methodology serves to inform youth that social injustices are constructs, not organic, and consequently, changeable.

> What perhaps distinguishes young people engaged in YPAR from the standard representations in critical youth studies is that their research is designed to contest and transform systems and institutions to produce greater justice—distributive justice, procedural justice, and what Iris Marion Young calls a justice of recognition, or respect. In short, YPAR is a formal resistance that leads to transformation—systematic and institutional change to promote social justice. YPAR teaches young people that conditions of injustice are produced, not natural; are designed to privilege and oppress; but are ultimately challengeable and thus changeable. (Cammarota and Fine 2008, 2)

This "formal resistance" is in direct contrast to the methodology of the "informal resistance" this book has been examining in the form of marches, strikes, and protests. These groundswell gatherings are very effective in raising awareness publicly, but rarely are directly accountable for progressive policy changes. Accountability for measurable and significant climate action seems to arise when direct "face-to-face" engagement occurs between the policymaker and the youth activist, not in a tokenistic way, but in a structure that desegregates youth and involves them in the policymaking process with the same respect of their perspectives and contributions that are customarily afforded to other marginalized groups (women, Black, Indigenous, and people of color (BIPOC), the impoverished, etc.).

Related to YPAR is a framework that takes closer aim at direct engagement: Participatory Action Research (Fals-Borda and Rahman 1991; McTaggart 1997; Selener 1997). In this approach, commonly referred to as PAR, the researcher is not a lone investigator but individuals in a collective. Together, or individually in the group, they are systematically addressing the same problem (climate change). Researchers engage in ongoing conversation and reflection with others, across generations, similarly poised to inquire and act. Research is therefore a collective process enriched by the multiple perspectives of several researchers working together. Second, the researcher, or more appropriately, researchers, are more or less "insiders" in a given situation. In other words, they are the stakeholders within a particular institution, government, corporation, organization, or community. For example, a PAR project in prisons would include prisoners as researchers (Cammarota and Fine 2008, 2). This approach makes all stakeholders accountable and therefore, all climate-related stakeholders should be part of the policy process: policymakers, the corporate sector, those marginalized communities most impacted by climate change (BIPOC, Indigenous communities), and, of course, youth. This participatory structure ensures representation and contributions from all parties impacted by climate change today and in the future and results in accountability for all in creating these progressive policies for their respective communities.

## 4.4 Evidence

When such productive and collaborative engagement takes place between the policymaker and the youth climate activist, any resulting policy, agreement, or program needs to be showcased, heralded, and documented publicly to provide evidence of success for this proposed collaboration. This needs to be done by the respective stakeholders involved in the process through their respective media outlets (mass and social media), inter-institutional publications, and academic publishing. "Doing so would help to provide a mandate for youth leadership in a variety of sectors and help counter misguided assumptions about the efficacy of youth input in shaping health priorities and action in local, national, and international arenas" (Spajic et al. 2019, 374).

Often, the invitation to participate through formal channels recognizes that the youth participants have formed an official group or organization that reflects similar areas of climate research and with similar intended goals. This is considered an acceptable alternative to conventional credentials. Sometimes, as well, stories told by youth on social media, and increasingly more through the medium of film, attract the attention of the policymaker providing "visible evidence"[1] of climate impacts. These stories reach the policymaker directly sent to them by youth as well as indirectly discovered by policymakers online.

Here are a few international examples of how youth activities outside the policymakers' offices opened the door to policy collaboration. Many of these activities come in the form of single-issue protests and award-winning short films made in the Young Reporters for the Environment's

[1] "Visible evidence" is a legal term used in documentary film scholarship to define factual representations of the truth. Link: https://www.visibleevidence.org/conference/visible-evidence-xxi-2/#:~:text=Visible%20Evidence%20is%20a%20%E2%80%9Ccollective, professionals%20involved%20in%20the%20study

annual competition and found in the UNFCCC's Youth Climate Report database resulted in evidence-based impact and new climate policy in national and local governments throughout the world:

## 4.5 CANADA: The Prime Minister's Youth Council

As mentioned previously in this chapter, Canada's Prime Minister's Youth Council is a bold initiative intended to involve its country's youth directly in policy discussions that impact on them. In an electronic correspondence with the Honourable Marci Ien, Minister for Women and Gender, Equality and Youth, she explains her dedication to consulting with youth in Canada's policy decisions:

> Across the country, young people have stepped up to break down systemic barriers, amplify diverse voices and create truly inclusive communities. They are actively sparking change to create a bright and progressive future. I want Canada's young people to know that I see them. I see their courage and how they are lifting up others, and I'll continue to keep them at the centre of our government's policy decisions.[2]

How does this commitment manifest itself in the Prime Minister's Youth Council? The Government of Canada actively engages with youth to help shape government priorities and actions to ensure that young voices are not only heard, but *respected*—an essential component of the Direct Approach which we will examine in detail in the next chapter.

In 2016, Prime Minister Justin Trudeau took on the additional role of Minister for Youth and established the Prime Minister's Youth Council to provide him with non-partisan advice, and to guide federal ministers and government officials on a wide range of issues that matter to them and their peers—issues such as promoting equality and protecting rights; enhancing supports for mental health; supporting Indigenous communities; and fighting climate change. Another important component of the Direct Approach—*Access*—is provided by this initiative.

In 2018, Canada launched a national conversation with youth. Young people discussed the issues that affect their lives, the types of supports they

[2] Received by Céline Marx, *PM's Youth Council – Book on Youth Climate Activism*, June 16, 2022.

need to succeed, and the ways they wish to be engaged. To facilitate *Access*, youth were encouraged to participate online, attend in-person roundtables, or host their own discussions and submit their ideas to the government.

Thousands of invitations were sent, directly or through youth-focused non-governmental organizations, to young people across the country to participate in engagement activities on topics such as accessibility legislation, climate change, agriculture, international relations, and youth employment. More than 5000 young Canadians took part in the national conversation, resulting in more than 10,000 individual responses and 68 submissions from youth-led discussions and youth-serving organizations ("Canada's Youth Policy" 2022). Partnerships with youth engagement experts and local youth-serving organizations ensured the government adopted a youth-centred engagement approach that resulted in the participation of young people with diverse identities, abilities, backgrounds, cultures, and economic and language profiles. Young Canadians then analyzed the responses from the national dialogue to shape key elements of the youth policy. This enterprise addresses the Direct Approach's need for *Collaboration*.

These consultations led to Canada's first Youth Policy, mentioned above. On May 2, 2019 the Prime Minister and Minister of Youth, Justin Trudeau, launched the Policy during the Canada Youth Summit. According to this Policy, young people have the right to influence decisions that affect them now and will affect them in the future the same way that other communities are consulted on decisions that affect them. In accordance with the Youth Policy, government officials from across Canada frequently consult the Prime Minister's Youth Council, their own departmental youth councils, youth across the country and organizations that serve youth as they work to reduce barriers and improve the lives of young people.

> The Prime Minister's Youth Council engages directly with the Prime Minister, members of his Cabinet and senior government officials several times a year, including in pre-budget consultations, in order to discuss issues of importance to them, their peers, their community and their country. This regularly results in the Prime Minister tasking ministers and federal departments and agencies to undertake further analysis of issues in order to address the Council's input. These budgetary considerations acknowledge the Direct Approach's requirement of *Finance* to this program.

### 4.5.1 *Canada's Youth Secretariat*

In addition, the Youth Secretariat serves as an information and resource hub for government departments, youth-serving organizations, and youth themselves. It supports a whole-of-government approach to addressing youth issues and provides advice to federal departments as they consider youth engagement and impacts of their programs, policies, and initiatives on young people. For example, the Youth Secretariat convened a Youth Advisory Group for the purpose of writing the first State of Youth Report, published in 2021.[3]

The group worked in consultation with the Prime Minister's Youth Council to give a voice to the priorities and the circumstances of the many youth across Canada who were engaged in the process. The group was composed of 13 young people from different backgrounds, genders, ethnicities, ages, and stages in life. The members analyzed and interpreted the data gathered during the engagement sessions to ensure that the perspectives of young people were heard and represented. With their names included in the State of the Youth Report,[4] the Direct Approach's important component of *Authorship* is satisfied.

As we will see in the next chapter, the Prime Minister's Youth Council serves as a model for adhering to all five requirements of the Direct Approach and its goal of youth engagement with policy participation.

## 4.6 MALTA: Plastic Bag Protest

Students from St. Thomas More College Zejtun Secondary School organised a march in the streets of their community to protest the use of plastic bags as part of their SUP (Stop Using Plastic) campaign. They also canvassed the Zejtun market, where they stopped consumers who were carrying products in single-use plastic bags and showed them alternative solutions. The media was present during the event, which helped the students share their message with a wider audience. The school gave out free reusable bags to those present at the event (Malta 2020). YRE organizers reported that this program was implemented by local government and that this protest with a proposed solution made policy changes and local shops introduced mesh bags instead of plastic. A small victory in the global

[3] https://www.canada.ca/en/canadian-heritage/campaigns/state-youth/report.html

[4] https://www.canada.ca/en/canadian-heritage/campaigns/state-youth/report.html#a11a

battle to end climate change, but one that provides evidence that youth climate activism can be effective when using the Direct Approach and appealing directly to policymakers.

## 4.7 CHINA: Working with YOUNGO

Xuan Zihan, a youth climate activist from China, has focused his activism on "high-level dignitaries and climate negotiators" at the United Nations through its official youth contingency, YOUNGO. He works to "ensure that youth climate demands and perspectives are meaningfully incorporated into climate governance" (Zihan 2022), one of the key objectives of YOUNGO. Zihan created YOUNGO policy last year through his contribution as a co-author of YOUNGO's official *Global Youth Statement* released at COP16 and COP26 in Glasgow, Scotland in 2021 and representing the contribution of more than 40,000 youth climate activists around the world. "We need to engage changemakers who have the ability and mandate to make a difference and hold them accountable for their commitments. The COP26 *Global Youth Statement* serves to do exactly this" (Zihan 2022). This type of global policy introduces the contribution of youth on climate issues for all world governments. Being released at COP26 sends a message to all world leaders that the global community of youth have not only spoken and been heard, but have contributed directly to meaningful climate policy that can be referenced in adapting its resolutions with youth on multinational levels.

While Zihan acknowledges the value of climate protests for youth, such as the Fridays for Future marches, he believes the more expedient path to significant change on a global basis will come from direct youth engagement with the policymaker.

> I see such protests as playing an integral role in adding pressure to policymakers and other stakeholders to prioritise climate justice and intergenerational equity in their pursuit of the objectives set out by the Paris Agreement and the Glasgow Climate Pact. Further, youth climate activism also aids in addressing eco-anxiety and provides an invigorating means for young people to enhance their morale by uniting for a common cause. That said, I believe that achieving significant changes in the system need to go beyond that. Specifically, young people must leverage on education and capacity-building to ensure that they have the knowledge and skills to contribute to fighting climate change. Furthermore, young people need to converse and seek

> partnerships with many other societal stakeholders (e.g. government, private sector, civil society), in order to come up with tangible actions and create systemic change (Zihan 2022).

His *Global Youth Statement* echoes this sentiment. In a section entitled "Underrepresented Groups", the statement calls for strengthening and capacity building for youth from the Global South, BIPOC, and other vulnerable and marginalized youth to ensure real representation. "We request that the floor at the COP Plenary and each negotiation stage has at least one person from the invisible BIPOC, majority of each country" (Lakhani et al. 2021, 12). The statement also demands that all youth be given a seat at the table by empowering them to participate directly in "all decision-making levels, both internationally and nationally" (Lakhani et al. 2021, 12).

To address the barriers frequently cited that prevent youth participation in policy discussions—finances, travel, presence—the statement creatively calls for the establishment of funds and the adoption of new technologies to connect youth who could not otherwise physically attend the meetings.

- "We request targeted scholarships, financial and travel support to ensure at least 1500 BIPOC, vulnerable, and marginalized youth participants in the next COP and 500 in the next Pre-COP".
- "We urge the use of technology and media to scale up the wide-spread of messages, attendance, collection of views in surveys, forums, workshops among others to ease the access to all youth".
- "We call for UN institutions to serve as means to assure connectivity to UN events, such as webinars, workshops, virtual meetings and conferences" (Lakhani et al. 2021, 12).

These demands take into account that scholarships and similar funding programs may be slow to emerge or never materialize, so when travel and accommodation at major policymaking events like the UN's annual climate summits is not possible, digital connectivity can be used to give youth from both the Global North and Global South a virtual seat at the table.

## 4.8 SOUTH AFRICA: THE JOHANNESBURG YOUTH CLIMATE ACTION PLAN (JYCAP)

In the past decade, several youth climate programs have emerged in South Africa to answer the call of youth for government climate action. Arising from the educational sector, students at schools and universities have formed these groups and organizations specifically to address the climate-related issues of biodiversity conservation, alien invasive species, reliance on fossil fuels, urban smog, drought, erosion, industrial pollution, and land degradation (Rodríguez-Labajos et al. 2019). The Johannesburg Youth Climate Action Plan (JYCAP) arose from a youth dialogue with the Department of Environment, Forestry and Fisheries (DEFF). In 2020, youth-targeted events with the DEFF minister were established with government and civil society to take place after school and to support youth transport and costs. This addresses one of the concerns expressed earlier regarding availability and costs that usually affect youth policy participation. Starting first on a municipal level, the City of Johannesburg launched the program. It was designed to be a *mind shift* and *practice shift* (Vogel et al. 2022, 5) in policy creation involving youth in a collaborative approach.

> Using such an approach youth are not only seen as being beneficiaries of environmental education programmes, many having their origins in the global north, but youth are also being involved directly, at the very outset of the policy creation and educational and social learning processes. Traditionally the route for policy science engagement is for youth to be called on in climate change 'matters of concern' usually set by an international funding or local funding agency and/or department, so that a 'box' can be ticked at the end of a so-called dialogue and/or action-based engagement (Vogel et al. 2022, 5).

The municipal experiment proved success enough to roll out a national version called the South African Youth Action Plan building on the positive outcomes and incorporating lessons learned from the JYCAP model. The national framework is comprised of five core pillars:

- Intersectionality
- Advocacy and Activism for Climate Action
- Good Governance

- Systemic Change
- Environmental Sustainability (SAYCAP 2021, 2).

### *4.8.1 Intersectionality*

In recognizing the complexity of the climate change crisis worldwide, this first pillar is defined by an intersectional representation of stakeholders: race, gender, disability, geographical location, access, income levels and other socio-economic conditions, compound the climate context and lived experience, particularly for people in under-resourced communities (SAYCAP 2021, 2). Youth's involvement in this program, as demonstrated through this pillar, ensures inclusion of all groups of youth, not just those privileged with accessibility and availability to participate in a national climate policy creation opportunity.

### *4.8.2 Advocacy and Activism for Climate Action*

Recognizing the Groundswell Approach the global community of youth are now using to advocate for climate justice, the South African government is encouraging the use of this mobilizing for the purpose of informing and raising awareness of key climate issues identified as crucial to youth. By harnessing this energy and its messages as a policy resource, this pillar identifies six components:

- Promote advocacy in schools and higher education institutions;
- Educate and support youth on how to engage with policy processes meaningfully and actively;
- Empower and capacitate youth activists with the skills and resources to engage with their communities on climate change and the socio-economic issues that intersect it;
- Support training and workshops that improve digital literacy and reduce the digital divide in vulnerable and low-income communities;
- Include youth at all stages of policy development and within decision-making spaces to foster a relationship with youth that encourages continuous and meaningful participation; and
- Lobby and support citizens to participate in government research and mitigation efforts that have a focus on climate related issues through explaining to them how climate change is interlinked with widespread community issues (SAYCAP 2021, 5).

The first two modules address education in a unique way. They encourage teaching and learning about the policy process, presumably assuming they are receiving their climate education both inside and outside the classroom already. This becomes important in addressing the complaint that youth activists who are too young to vote are unaware of the policy process. Through this curriculum, youth become empowered to be valuable contributors when sitting at the table with the policymaker.

### 4.8.3 *Good Governance*

The third module enhances the ability of youth to advocate climate action already practised through their Groundswell Approach to marches and protests by providing them with the "skills and recourses" for more significant community engagement—an important quality to bring to the table in policy discussions.

### 4.8.4 *Systemic Change*

The fourth module acknowledges that certain communities of youth in South Africa may be remotely located and not have access to WiFi or digital communication devices, otherwise known as the "digital divide". This is an aspect of youth climate activism that is often overlooked, but one addressed by certain global programs such as York University's Planetary Health Film Lab and the UNFCCC's Youth Climate Report. These programs seek out remote communities, often Indigenous, and provide the same "training and workshops" to narrow this divide and give voice to youth communities and their unique relationships with climate impacts.

### 4.8.5 *Environmental Sustainability*

The fifth pillar is perhaps the most important in relation to youth's direct participation in climate change policy creation. By including youth at "all stages" of policy development and within "decision-making" spaces, they become empowered to draft legislation based on their community's needs. It also ensures that policy created within this framework is crafted with the future of the community in mind. With this commitment, youth not only have a seat at the table, but they are actively writing policy that represents the intersectionality of youth communities throughout South Africa. The success of this approach can serve as a model for other national entities and non-state actors to follow.

The last pillar sounds like another version of education and outreach, but this one is aimed not at youth communities in particular, but all national citizens to participate in research and mitigation programs and strategies that implement informed programs that address the current and future impacts of climate change. This is the “action” Greta Thunberg and her followers so often demand.

On paper and in theory, this framework seems to address all the obstacles usually identified as holding youth back from meaningful participation with governments in creating progressive climate policies. But how effective is this in practice? How do youth respond to the tasks and responsibilities set before them? One 18-year-old student in South Africa, Khahliso Myataza, who now holds the title Youth Policy Committee member within the South African Institute of International Affairs, recounts her experience with the program:

> (When) I was in Grade 11, I joined the lead drafting team for the Johannesburg Youth Climate Action Plan (JYCAP), which taught me how to be more intentional with my advocacy and my voice. Even though I was still in high school, I was working with university scholars and it was intimidating. It took a lot of effort and conscious self-reassurance to get over my feeling of being an imposter – who was I to make these statements, wasn’t I too young to contribute, what did I really know? Many young people suffer from this kind of ‘imposter syndrome’ and it’s hard to find your place. But being on the drafting team really helped me to grow – I began to understand the importance of the work we were doing, about how to make contributions and be open to critique and criticism, and it pushed me to think and develop in so many ways. It was through this experience that I came to realise how important it is for young people, regardless of their age, to be involved in policymaking processes. Going through the Johannesburg YCAP drafting process was tiring yet exhilarating. We started by gathering input from young people across the different regions of Johannesburg. We also held a dialogue on climate change and climate activism to conceptualise the ‘Johannesburg’ we wanted to live in – what did it look like and how could we get there? Participants spoke about intersectionality, decolonisation, the just transition, decentralisation, and other issues pertinent to climate change. These became the pillars that anchored the document. The team would spend hours working on and perfecting a single section, and even though it was exhausting, we knew we were doing something important. Everyone was passionate about climate change and climate justice and

> we were not deterred, even when it was difficult. We knew this cause was bigger than all of us and that we had an important story to tell. Being in these policymaking spaces gave me a glimpse into the world of representing others and it made me realise how important policies are – they impact our whole city's trajectory. We had to ask ourselves, are we going to be the city that ignores climate change, or are we going to be the city that works earnestly to hold big corporations accountable and decrease our carbon footprint? Policies only work for the population if that population is represented in the negotiating rooms, and I really began to understand the importance of including young people within that space. As the continent with the largest youth population, it is imperative that young people are engaged in policymaking processes. We are the future that will inherit the continent and we need to take a holistic approach, one that is intersectional, decolonised and evolving, to ensure we bring the greatest benefits to the whole population. (SAYCAP 2021, 7–8).

This report candidly expressed the student's initial anxiety and imposter syndrome that many young people feel when introduced into an unfamiliar environment. She speaks of the "tiring" and "exhausting" process involved in creating the pillars we examined earlier in this chapter, but at the same time, she says the exercise was "exhilarating" because she knew how important her work was not just to youth but to the intersectional, decolonised, and evolving population as a whole. This first-hand report reveals that the goals of the pillars within the South African Youth Climate Action Plan (SAYCAP) youth engagement framework have been met. She expresses a clear understanding of intersectionality and its importance in any climate policy now for future generations. She also expresses some surprise at learning why youth participation is important revealing a new understanding of the policy process previously not known.

## 4.9 The Netherlands: Theoretical Framework for Youth Climate Policy Participation

In a 2019 study on the underrepresentation of youth in Dutch climate politics, Willemijn Schreuder provides a theoretical framework for their participation in climate policy discussion and, most importantly, creation. As with any other marginalized community, youth have an experience and perspective that is unique and often unknown or misunderstood by those in power leading to policy that either ignores them and their needs or is not created altogether.

Schreuder concludes that climate policy must include youth participation to ensure an increased capacity of the next generations to respond effectively to the unpredictable futures of climate change (Schreuder 2019, 6). In identifying the constraints that stand in the way of meaningful policy participation for youth, the study cites the research of Barry Checkoway who refers to the lack of credibility in science knowledge and policy understanding this chapter has previously examined, using the term "adultism", a phenomenon which means that adults see themselves as better than young people because of their age (Checkoway, 2011). But it's not just the adults who believe that today's youth are incapable of consequential collaboration; it is also the youth themselves who believe they might be in over their heads when it comes to policy creation.

To formalize a structure of youth participation that considers these constraints on both sides of the adults and the youth, a Dutch youth movement known as *de Jonge Klimaatbeweging* (JKB) was established to not only influence climate and sustainability policy by means of campaigns but also and, more significantly, through conversations with politicians.

> After the climate agreement in Paris in 2015, several YOUNGO (constituents) in the Netherlands felt that young people were still not represented in Dutch politics. ....With the founding of JKB, a lobby document was made along with 40 youth organisations as a manifesto which was presented towards the government. This manifesto eventually turned into the young climate agenda. Which is a vision document of 40 youth organisations with their ideas about 2050 and it acts as lobby document of the raison d'être of JKB (Schreuder 2019, 10).

As another manifestation of the Groundswell Approach, large gatherings of youth climate *groups*, as opposed to individuals, demonstrate to the government that the constraints and "adultism" prejudiciously imposed on youth is countered through this collective of organized parties. This particular grouping intends to formalize youth participation, and secure a seat at the table, in national climate policy creation, and not just influence. Their manifesto serves as a resource of the perspectives and demands of the youth community so government officials can incorporate them into national climate policy.

On their website, the JKB describes itself as an organization that "is based on a clear idea: young people deserve a voice to shape their own sustainable future. We are the generation that will have to live with the consequences of climate change. Today's decisions determine what the world will look like tomorrow…By entering into discussions on behalf of our supporters with politicians, policymakers and the business community, we give young people a voice in their own future" ("About Us", *jongeklimaatbeweging.nl* 2022). The success of this ideology has spawned another manifestation of the Groundswell Approach as mutual interest has spread the JKB globally. In addition to the collective of youth organizations in the Netherlands, the JKB lists partnership with other countries following their model: Uganda, Bangladesh, Chile, Egypt, Mexico, Nigeria, South Africa, India, and Qatar ("Global Partnerships", *jongeklimaatbeweging.nl* 2022).

To further inform the policymaker of the interests and perspectives of the global community of youth, the JKB develops a series of "position papers" to aid in the creation of new climate policy in the Netherlands. These position papers are also used by their partner country members to assist their own national climate policymakers. Many of these papers not only present the position of youth on several climate and sustainability issues, but also offer concrete proposals to do so. Some of the areas these papers address include the economy (The Circular Economy for Today and Tomorrow), health (Climate Change and Public Health), and housing (Climate, Housing, and Youth Participation). One paper on nuclear energy participates in a policymaking roundtable discussion by being presented to the House of Representatives in advance of the meeting so it would be tabled and used as a resource ("Position Papers", *jongeklimaatbeweging.nl* 2022).

This framework of manifestos and position papers do not simply represent a general community of Dutch youth, but an informed community of youth through its 40 formal organizations of youth climate activists. Their scientific research and societal demands are submitted directly to policymakers to guide their climate policy creations and, when necessary, direct collaboration between the two parties ensures mutual understanding and leads to new climate policy that accurately represents the country's community of young people who will be implementing these laws into their future lives.

## 4.10 The United Nations: YOUNGO

The last national model of youth engagement with climate policymakers this chapter will examine is multinational: YOUNGO, the official youth constituency of the United Nations. After an initial request from youth for official standing with the UNFCCC at its COP conferences, YOUNGO was created in 2009 and officially confirmed in 2011 to formally recognize the global community of youth at the COP conferences by the United Nations Framework Convention on Climate Change. The year-long process eventually recognized youth as an observer constituency to UNFCCC. While YOUNGO has an official status with the UNFCCC, it is still an entity that is youth-led and self-organized ("History" 2022). YOUNGO focuses on the following areas:

1. Awareness, Knowledge, and Capacity Building
2. Collaboration, Cooperation, and Network
3. Policy, Lobby, and Advocacy
4. Youth Action ("YOUNGO" 2022)

According to the UN's website, YOUNGO "representatives make official statements, provide technical and policy inputs to negotiations and engage with decision-makers at the UN climate change conferences, and promote youth participation in climate change projects at local and national levels" and it comprises 200 youth NGOs and more than 5500 individual members (YOUNGO 2022). This may seem that they do indeed have a valuable seat at the table, but many believe its contributions fall short of actual policy creation collaboration. In their own *Global Youth Statement* released in 2021, YOUNGO's authors repeatedly call for more direct participation in the policymaking process. In the section titled "Top Policy Demands" the paper specifically calls for increased policy involvement:

> Our overarching demand that unifies all themes is that the youth should be actively and meaningfully included in all decision-making processes concerning climate change governance and implementation. We call for these policy demands to be integrated into national and international agendas and commitments. We demand an intersectional approach to youth inclusion in environmental governance.... (Lakhani et al. 2021, 2)

There is not a lot of scholarship on youth engagement with the United Nations through its constituency YOUNGO, though Harriet Thew, a Research Fellow with the School of Earth and Environment at the University of Leeds has focused a lot of her work on this. She believes that one of the obstacles YOUNGO faces in consistent and meaningful engagement with UN policymakers comes from the high turnover of its membership. This occurs when young members grow beyond the age of youth and move on. "YOUNGO's high turnover makes it difficult for youth to be held accountable for implementing governance tasks and, as a result, authority is not readily delegated to them. A double-edged sword, this also prevents them from holding others accountable for previous conversations and commitments" (Thew et al. 2021, 881). The study she conducted with Lucie Middlemiss and Jouni Paavolva, *Does youth participation increase the democratic legitimacy of UNFCCC-orchestrated global climate change governance?* proposes a solution: "UNFCCC orchestrated initiatives could support young people to overcome these challenges by offering regularly scheduled opportunities for input which could be prepared for in advance, and formally documented to enable newcomers to hold decision-makers accountable for issues raised in previous meetings" (Thew et al. 2021, 881–882).

Other scholarship suggests that though the observer group may have its flaws in maintaining consistent membership, its unique status with the UNFCCC provides opportunities for meaningful policy engagement other youth groups do not have: "Despite their observer status, (YOUNGO is) advancing various programs to augment young people's participation and attempt to influence international negotiations and conferences, including through the organization of annual Conferences of Youth (COY). From our experience, the engagement of young people at these negotiations has amplified advocacy efforts and put pressure on states to meet their commitments on emission reductions" (Gasparri et al. 2021, 98). YOUNGO is just a little more than ten years old and while they may not have a seat at the table in terms of actual policy participation, they do hold an important place in the UNFCCC, a foot in the door that allows them opportunity to learn how to participate and with this access to achieve this goal.

## 4.11 A Proposed Framework for the Groundswell Approach

From the research presented in the previous chapters, we can construct a proposed framework for successful youth climate activism using the Groundswell Approach. The Fridays for Freedom model offers a valuable structure of mobilizing information, event promotion, participant engagement, public pressure on governments, and media outreach for organizing protest marches and gatherings. The Internet in general, social media in particular, and other digital platforms like websites and posts all contribute to form a dynamic ecosystem of event promotion, participant engagement, and event documentation. Organizers of such events must be aware of their local restrictions on public gatherings and secure the necessary permits and security required by their municipal, regional, or federal governments. This component of our framework is called the *Protest Protocol*.

We have seen the importance of leadership, both for the movement and for the young people who assume the role. Reaching out to existing youth climate leaders to strategize participation in high-profile events is always important, but within this framework, a collaboration of message and strategy aimed directly at the policymaker may yield positive results. Youth climate leaders need to speak at rallies, marches, conferences, and other media-attended events to bolster their profile locally and nationally. If an area does not have a youth climate leader, put a social media call out to find one. It is important to arrange meetings at this stage with local government officials and these young leaders. Opening a dialogue is the first step in establishing youth participation in policy creation as we will see in the chapters to come. This component of our framework is called the *Leadership*.

Ultimately, government climate action and policy related to climate justice is the goal of youth climate activism. We have seen how effective protest marches like Fridays for Future can be around the world, especially when coordinated to take place on the same day. The local march becomes a global event. The same strategy can be applied on a global basis to a series of events that reach out to government officials directly. As we will see in Chap. 6, a phone zap is a new method of concentrating contact with policymakers and keeping them accountable through social media posts at the same time. Done on a coordinated basis like the Fridays for Freedom marches, the collective voices of youth climate activists are amplified in a targeted way to those in power who are ultimately responsible for the policy that is required to end climate change. This component of our framework is called the *Action*.

Once contact has been made with the policymaker, the goal of the first meeting is to secure a second meeting. Instead of demanding climate action, which can be a little vague and aggressive to government officials at this introductory meeting, the establishment of a Youth Council within the government should be proposed. This recognizes what many politicians realize is a large and important, yet often overlooked, sector of the people they govern so they are likely to agree. This gives the youth climate activist a meaningful foot in the door. From this internal position, participation in policy discussions, negotiations, and creation is more likely to occur. With this privileged position comes the responsibility of knowing climate data and policy procedure. Direct experience in these meetings will provide some of that knowledge, but the youth climate activist must expand their understanding of each to make meaningful contributions. Empathy with the policymaker is also a key component often overlooked by youth climate activists. Try to understand the challenges they face in advancing progressive climate policy and offer assistance in overcoming these obstacles. Local climate leaders can integrate this strategy with other climate leaders around the world through a social media campaign that stresses involvement in government environmental policy sessions. This component of our framework is called the *Nexus.*

The **PLAN** framework—**P**rotest Protocol, **L**eadership, **A**ction, and **N**exus—applies the Groundswell Approach and grows the protest movement beyond the FFF marches and keeps the pressure on government to act with the urgency the climate crisis demands. It also expands the participation of youth and others at the policy level to magnify the global demand for change. With dynamic leadership in place worldwide, representatives of the movement can deliver similar messages with the same talking points about climate change impacts and solutions. This is already being done to some degree, but within this framework, the aim is to establish representation within government to contribute to policy on a regular basis. This presents a solidarity among the world's youth climate activists and strengthens their position to conventional media and government officials. Another groundswell activity made with a concerted effort on a designated day throughout the world is PLAN's reach-out to local environmental policymakers. With the goal of getting a second meeting, the first meeting is usually a success, paving the way to establish a seat at the table through the establishment of a Youth Council. Once in this position, participating in environmental, sustainability, and climate policy meetings becomes much more productive and, when instigated worldwide, more globally impactful.

As we will see in the second half of this book, a paradigm shift in how youth climate activists perceive the policymaker—from reluctant adversary to collaborative partner—is proposed to accelerate the path to progressive climate policy. The two frameworks proposed in this book—the PLAN model and the Direct Approach strategy—provide a roadmap to this destination.

## 4.12 Conclusion

The proposed framework for the Groundswell Approach – the PLAN Model – and the proposed framework for the Direct Approach detailed in the next chapter, are both intended to establish a mutually beneficial relationship between the state and the global community of youth.

Ideally, the global community of youth should be treated with the same respect as policymakers at all levels of governments afford other marginalized communities, such as Indigenous Peoples, LGBTQ+ members, and racially diverse groups. These populations are consulted as a matter of due process in most policy creation because they are seen to be underrepresented communities; however, many believe these groups are not participating enough (Grosse and Mark 2020). Governments and their policymakers make efforts to be inclusive of these communities to ensure their needs, cultures, and perspectives form part of the policy that affects all people. So, too, then should the global community of youth be regarded in a similar way and be included in all climate policy creation processes that will directly impact on them now and in the future they will soon inherit.

## Works Cited

"About the Council", *Prime Minister's Youth Council.* Accessed: April 24, 2022. Link: https://www.canada.ca/en/campaign/prime-ministers-youth-council/about.html.

"About Us", *jongeklimaatbeweging.nl.* Accessed: May 1, 2022. Link: https://www.jongeklimaatbeweging.nl/over-ons.

"Canada's Youth Policy", *canada.ca.* Accessed: June 16, 2022. Link: https://www.canada.ca/en/youth/programs/policy.html.

"Global Partnerships", *jongeklimaatbeweging.nl.* Accessed: May 1, 2022. Link: https://www.jongeklimaatbeweging.nl/project/global-partnership.

"History", *YOUNGO.* Accessed: May 4, 2022. Link: http://www.youngo.uno/about/history.

Kenya, "National Climate Change Framework Policy", *Ministry of Environment and Natural Resources*, Kenya: May 2017. Accessed: April 25, 2022. Link: http://www.environment.go.ke/wp-content/uploads/2018/08/Climate-Change-Framework-PolicyMay2017.pdf.

"Malta", Young Reporters for the Environment, 2020. Accessed: April 30, 2022. Link:https://www.yre.global/stories-from-countries-blog/2020/7/2/malta.

OECD: *Evidence-based Policy Making for Youth Well-being: A Toolkit*, OECD Development Policy Tools, OECD Publishing, Paris, 2017. https://doi.org/10.1787/9789264283923-en.

"Position Papers", *jongeklimaatbeweging.nl*. Accessed: May 1, 2022. Link: https://www.jongeklimaatbeweging.nl/position-papers.

SAYCAP (The South African Youth Climate Action Plan), 2021. Accessed April 30, 2022. Link: https://saiia.org.za/wp-content/uploads/2021/10/The-South-African-Youth-Climate-Action-Plan-2021.pdf.

"YOUNGO", *United Nations Climate Change*. Accessed: May 4, 2022. Link: https://unfccc.int/topics/education-youth/youth/youngo#eq-1.

Benkenstein, Alex, Romy Chevallier, Desiree Kosciulek, Ditebogo Lebea, and Kiara Worth. "Youth Climate Action and the Role of Government", *Youth Climate Advocacy*. South African Institute of International Affairs, 2020. Accessed: April 25, 2022. http://www.jstor.org/stable/resrep29503.

Brooks, Spirit D., Steven M. Braun, and Dan Prince. "Critical Consciousness in High School Outdoor Experiential Environmental Education." *Journal of Experiential Education* (2022): 10538259211068800.

Cammarota, Julio, and Michelle Fine. *Revolutionizing Education: Youth Participatory Action Research in Motion*. New York, NY: Routledge, 2008.

Checkoway, Barry. (2011). "What is youth participation?", *Children and Youth Services Review*, 33(2), 340–345.

Checkoway, Barry, Tanene Allison, and Colleen Montoya. "Youth participation in public policy at the municipal level." *Children and youth services review* 27.10 (2005): 1149–1162.

Erickson, Jessica. "Experiential Education in the Lecture Hall." *NEULJ* 6, 2013, 87.

Fals-Borda, Orlando, and Anisur Rahman (1991). *Action and Knowledge: Breaking the Monopoly with Participatory Action-Research*. New York: Apex Press; London: Intermediate Technology Publications.

Gasparri, G., Omrani, O. E., Hinton, R., Imbago, D., Lakhani, H., Mohan, A., Yeung, W., & Bustreo, F. (2021). "Children, Adolescents, and Youth Pioneering a Human Rights-Based Approach to Climate Change", *Health and Human Rights*, *23*(2), 95–108.

Grosse, Corrie, and Brigid Mark. "A Colonized COP: Indigenous Exclusion and Youth Climate Justice Activism at the United Nations Climate Change Negotiations", *From Student Strikes to the Extinction Rebellion*. Editor: Benjamin J. Richardson. Northampton, MA: Edward Elgar Publishing, 2020.

Han, Heejin, and Sang Wuk Ahn. "Youth Mobilization to Stop Global Climate Change: Narratives and Impact." *Sustainability* 12.10, 2020.

Harriet Thew, Lucie Middlemiss & Jouni Paavola (2021) "Does youth participation increase the democratic legitimacy of UNFCCC-orchestrated global climate change governance?", *Environmental Politics*, 30:6, 873–894, DOI: https://doi.org/10.1080/09644016.2020.1868838.

Henderson, Karla A. "Got research in experiential education? Theory and evidence." *Journal of Experiential Education*, Volume 26, No. 3, 2004, 184–189.

Lakhani, Heeta et al. *The Global Youth Statement*, Glasgow: YOUNGO, October, 2021.

McTaggart, Robin (1997). *Participatory Action Research: International Contexts and Consequences.* Albany: State University of New York Press.

Rodríguez-Labajos, B., I. Yánez, P. Bond, L. Greyl, S. Munguti, G. U. Ojo, and W. Overbeek. 2019. "Not So Natural an Alliance? Degrowth and Environmental Justice Movements in the Global South." *Ecological Economics* 157: 175–184.

Schreuder, Willemijn. *Youth Participation in Climate Change Policy-making.* Doctoral Dissertation, University of Groningen, Netherlands. 2019.

Selener, D. (1997). *Participatory Action Research and Social Change.* Ithaca, NY: Cornell Participatory Action Research Network.

Spajic, Luke & Behrens, Georgia & Gralak, Sylvia & Moseley, Genevieve & Lindolm, Daniel. (2019). "Beyond Tokenism: Meaningful Youth Engagement in Planetary Health", *The Lancet Planetary Health*, e373–e375. https://doi.org/10.1016/S2542-5196(19)30172-X. Accessed: April 24, 2022. Link: https://www.thelancet.com/journals/lanplh/article/PIIS2542-5196(19%2930172-X/fulltext.

Teixeira, Samantha, Astraea Augsberger, Katie Richards-Schuster, and Linda Sprague Martinez. "Participatory Research Approaches with Youth: Ethics, Engagement, and Meaningful Action," *American Journal of Community Psychology* 68, no. 1–2 (2021): 142–153.

Vogel, C., Nkrumah, B., Kosciulek, D., Lebea, D., Booth, T., & Brown, M. (2022). 'Empowering Youth as Change Agents for Climate Change in South Africa': Challenges, Caveats and Course Corrections. *Journal of Youth Studies*, 1–21.

Zihan, Xuan. Electronic Interview, conducted by Mark Terry. April 29, 2022.

PART II

# The Direct Approach

Indigenous Inuit youth at a COP25 press conference in the Blue Zone. Meeting with UN policymakers to inform and influence new climate policy from their unique perspectives. (Photo Credit: Youth Climate Report © 2019)

CHAPTER 5

# The Direct Approach: A Participatory Framework for Meaningful Youth Engagement with UN Policymakers

The first half of this book investigated the historical and current ways in which youth have engaged with environmental activism through the lens of the Groundswell Approach. Before and after the digital revolution, youth around the world gathered in large numbers to protest the status quo that frequently ignored threats to the environment and, consequently, the global ecology for all life on the planet. From the physical gatherings at marches and school strikes to the virtual activism of social media protest campaigns, youth have found success with this approach in raising awareness of the climate change crisis in unprecedented numbers. However, we have also found that this approach is not necessarily successful in creating measurable change to existing environmental policies. Changes come in small measure, and rarely are they attributed to the protests of youth, regardless of how many voices are being raised.

The second half of this book will examine the more focussed method I call the *Direct Approach* that endeavours to provide a space and structure where meaningful dialogue and participatory engagement between youth and policymaker can take place. The goal of this approach is to establish a similar consultation that is often afforded marginalized communities by governments when considering new policy. These engagements aim to be inclusive of underrepresented communities to ensure that new legislation takes into account diverse cultures, distinctive socio-economic structures, and unique perspectives that may be overlooked by those in power.



M. Terry, *Speaking Youth to Power*, Palgrave Studies in Media and Environmental Communication,
https://doi.org/10.1007/978-3-031-14298-7_5

Providing youth this same affordance is not something often done by governments in general, and in the United Nations, such formal consultation is relatively new dating back to only 2009 ("History", 2022). Citing a general lack of interest and experience in policy creation, and a lack of credibility in climate science expertise, the policymaker has been reluctant to take this significant sector of the global population seriously. It is not without some irony that the population of people who will be most impacted by the lack of climate policies of today are those who are not currently being consulted.

The Direct Approach represents an amalgamation of previous models of youth participation theory informed by and for UN agencies, most notably, Roger A. Hart's "Ladder of Participation" (1992) prepared for and published by UNICEF; Harry Shier's "Pathways to Participation" (2001); and Helen Cahill and Babak Dadvand's "7 Ps of Participation" (2018). It should be noted here that several of these models refer to youth by the slightly archaic term of "children", but since they often use the same age definitions that current scholarship uses to define "youth", this examination will consider both terms to be interchangeable. This chapter will examine the strengths and weaknesses of these models to synthesize the elements that best recognize today's emerging relationship between youth and power in a methodology of policy participation with the UN that I propose as a new framework called the Direct Approach.

## 5.1 Hart's Ladder of Participation

Published by the International Child Development Centre of the United Nations Children's Fund (UNICEF) in 1992, Roger A. Hart's paper entitled *Children's Participation: From Tokenism to Citizenship* proposes one of the first frameworks for involving youth in the policy process. Based on Sherry R. Arnstein's "ladder of citizen participation" published in the *Journal of the American Institute of Planners* in 1969 (Arnstein 1969), Hart refers to his structure as a Ladder of Participation for youth and details the various forms of political engagement that youth currently experience. The higher up the ladder one goes, the more meaningful the engagement. The levels he suggests are:

1. Manipulation
2. Decoration
3. Tokenism

4. Assigned but informed
5. Consulted and informed
6. Adult-initiated, shared decisions with children (youth)
7. Child- (Youth-) initiated and directed
8. Child- (Youth-) initiated, shared decisions with adults (Hart, 8)

These are based on the premise that there are power relations between youth and those with whom they need to engage. "Young people's participation cannot be discussed without considering power relations and the struggle for equal rights. It is important that all young people have the opportunity to learn to participate in programmes which directly affect their lives" (Hart, 6).

### 5.1.1 *Manipulation*

The first rung of the ladder refers to how adults sometimes use children without explaining to them the issue in which they are participating, like carrying signs or wearing a messaged t-shirt in a climate protest march. Without understanding what they are doing, young people are being manipulated by adults. They are "participating", but not necessarily of their own free will and in the service of a cause with which they are unfamiliar. Other times, adults may ask youth for their contribution or feedback on an issue, but then do not offer the results of their analysis of these contributions. Decisions are made by the engaging adults on behalf of the youth and without their consultation. This is particularly evident in policy meetings where youth get to speak in a one-way discussion. The information and perspectives they provide may or may not be used by the policymaker, but in both scenarios, the youth is unaware of the impact and outcomes which may be used to forge a policy that is not necessarily in their best interests. In this case, there is no mutual respect in the process.

### 5.1.2 *Decoration*

Similar to manipulation, adults may use youth to show that their cause impacts on young people and that they have their implicit support, but again, youth participants may not be aware of the cause or have an understanding of it. Having youth participate in a group in costume and perform a song or dance at a climate event *decorates* the crowd assembled to send a message that the protest is in the best interests of youth. This may very well be the case, but without the intellectual and cognizant engagement of the youth involved, their meaningful participation is absent and any resulting policy is made without their consent or contribution.

### *5.1.3 Tokenism*

This rung of the ladder is seen most often in situations where those in power invite dialogue with youth, but have no intention of acting on anything they present. The opportunity to engage is usually prescribed by the adult power in a manner comfortable to them but not to the youth, such as an invitation to present a PowerPoint in a board room environment. Access may be granted to a privileged encounter with those in power, but with no consideration of the community of youth in terms of meeting structure or of the message and content presented, such participation is meaningless and simply *tokenistic.* The illusion is created that open dialogue has been established, and often the participating youth are gaslighted into believing they are making a difference, but in the end the same inaction prevails rendering the encounter an example of tokenism. It is a little higher on the ladder because while youth may not achieve the goals they seek, they can still come away with a valuable experience of how those in power work and can subsequently adopt different tactics for the next opportunity they have that can make the adults in the room more accountable.

### *5.1.4 Assigned But Informed*

This level is the first one that recognizes an informed youth in the participatory policy process. To be truly participatory, Hart suggests that four requirements of youth must be established:

1. The children understand the intentions of the project;
2. They know who made the decisions concerning their involvement and why;
3. They have a meaningful (rather than "decorative" or "tokenistic") role;
4. They volunteer for the project after the project was made clear to them (Hart, 11).

In this case, an assignment to participate may be made and all four requirements may be met, but if the opportunity to engage in discussions is not there, policy participation is not there. The example provided is when the United Nations invites youth to serve as pages at youth conferences at UN headquarters in New York. Access, again, is granted and they are informed of their duties and their role in this capacity which is also meaningful in light of the theme of the conference. They understand the intentions of their involvement and freely volunteer, but the experience, while valuable, is not directly contributory to the policy process.

### 5.1.5 Consulted and Informed

While youth are beginning to be consulted on climate policy more frequently, they are seldom seen as a source of information for the policymaker. There is also a need to consult the global community of youth in the same manner as other communities are consulted, like Indigenous Peoples, women, and the disabled. There is an implicit understanding that these communities possess cultures and perspectives not necessarily known to those outside of those communities and in particular, the policymaking community, therefore, consulting with them to ensure accurate representation is required. This achieves the goal of inclusivity the UN so often claims to embrace as an essential global policy component. However, we are not there yet, but we can see examples of how such consultation with youth might work in the corporate world. A videogame designer may recruit and invite a group of young people to consult on a new game to ensure the aspects of the game their customers desire are included. This is common in market research studies, but where this example fails in terms of mirroring the intended engagement with youth at a UN policy level, is the participating youth are not informed of the results of their consultation. When the videogame designer creates the game based on the input of the youth consultants and has them play it, the participants are now informed. The benefit to this step is clarity. What may have been described by the youth consultant, may have been misinterpreted by the game designer. By having the consultant play the game and identify the misinterpretation, the game designer can then fix the problem and create a game (policy) that is truly representative of their customer and one that is equally desired by the customer. This model takes into account that even a professional in the field may not know what a large sector of their audience (constituency) wants or needs and when that information is provided, they go that extra mile to ensure accurate applications of that information are made. This is an important rung of the ladder that is beginning to take place in some national governments, but not to the same extent at the multinational level.

### 5.1.6 *Adult-Initiated, Shared Decisions with Children (Youth)*

This structure is referred to as "true" participation for youth since they are directly involved in the decision-making process, even if the policy is initiated by adults. In this political environment, youth are consulted as the policy being created directly involves them. The example here is a municipal playground or skateboarding park. Young people are the ones intended to use this public space so it makes sense to seek their input on desired rides and ramps. They can also provide safety perspectives by informing the adult policymaker which components of the park may present hazards to them. They can also advise on which components to include and which ones to dismiss as they will be the users of this equipment. In the end, youth make a valuable and informed decision on the design of these public spaces involving them directly and meaningfully in the policy process.

In the arena of climate policy, this format of engagement is extremely valuable as the policy proposed will not only impact on young people today, but will also likely be in place when they become adults. Making decisions that impact only on current adult scenarios and preferences is short-sighted and does not take into account a vast majority of the global population that youth represents.

### 5.1.7 *Child- (Youth-) Initiated and Directed*

These situations are rare, but ideal in terms of representation. Instead of an adult proposing a policy or project with adult perspectives and restrictions that may preclude certain contributions from youth participants, this structure begins with youth and is directed by them as well. Canada's York University has an eco-campus in Costa Rica called Las Nubes (The Clouds). Here students learn about environmental issues through a progressive form of experiential education. One of the projects identified by the students in the early years of the campus was to plant trees throughout the grounds of the campus. The youth in this example initiate the project and source the saplings to be planted. When the young trees arrive, students attending the campus each term plant these trees where they believe would be best. Faculty and staff are available for support, but this serves as a good example of an environmental project initiated and directed by youth.

A similar youth-led climate project is in place with the UNFCCC. The Youth Climate Report began as a project to inform delegates attending the COP conferences of climate research, impacts, and proposed solutions throughout the world. The method of reporting—video stories—was initiated by youth and all projects are made exclusively by youth aged 18–30. Administrators of the database are adults, but they do not contribute in the artistic representation or messages of the reports leaving the youth participants to express themselves according to their own facility with digital media production and respective national cultural influences. While these youth are not "seated at the table" of policy creation, this youth-initiated and directed project serves to inform and influence the environmental policymakers of the United Nations.

### 5.1.8 Child- (Youth-) Initiated, Shared Decisions with Adults

The rarest of all forms of participation for youth in policy creation projects, according to Hart, are the ones in which youth initiate the research and share the results with decision-making adults, not only informing them, but participating in the final decisions that ultimately result in policy. This is rare because adults are all but absent from the process leaving the methodologies of the research and the subsequent findings to be suspect by adults charged with incorporating this data. One successful example of this is the act of bird-watching in which many youth engage.

A citizen science project known as *eBird*[1] encourages observations of bird species around the world from non-scientific participants, referred to as "birders". These birders have no age range meaning that any youth who wishes to initiate a bird-watching activity, alone or in groups of other youths such as classmates, can participate and contribute data. The only adult involvement comes in the form of a questionnaire provided to birders to complete as a way of cataloguing the data, usually in terms of location, season, time of day, and so on. The program also considers youth participants by making accessibility to it through a mobile app, a technology most youth today are not only familiar with, but one they prefer to use. The submitted results are in a form that the scientific community can easily

[1] https://ebird.org

codify and add to a GIS database which informs policy related to research on migratory patterns and possible anthropogenic reasons for changes in habitat and population numbers. The youth-friendly project, known as the "world's largest birding community" (*About eBird*, 2022), may not directly involve youth in the eventual policymaking decisions, but their citizen-scientific contributions to those decisions are clearly involving them and making their voices heard in a meaningful manner.

Collectively, these ladder rungs provide a solid foundation upon which to build a framework that takes into account today's digital world. Many more points of access and communications networks exist today that were not in place when Hart proposed this model for UNICEF. Of particular value is the emphasis he places on youth "initiation", a presumption that any meaningful participation should start with and be directed by youth instead of the adults in positions of power. There are obvious shortcomings of such a practice which this book has addressed with respect to the "expert" status and "experienced" background many youth are missing due to their age, but philosophically, the concept of youth initiation and direction in policy participation ensures representation of the community's goals, perspectives, and culture.

## 5.2 Pathways to Participation

The second model of policy participation for youth was created in 2001 by Harry Shier based on the guidelines provided in Article 12 of the United Nations Convention on the Rights of the Child of 1990. This UN branch, like UNESCO in the model before, focuses on youth, children, and their rights. In particular, Article 12 makes a clear declaration that youth must be consulted with and participate in policy discussions that impact directly on them. Here are the two clauses of Article 12:

1. States Parties[2] shall assure to the child who is capable of forming his or her own views the right to express those views freely in all matters affecting the child, the views of the child being given due weight in accordance with the age and maturity of the child.
2. For this purpose, the child shall in particular be provided the opportunity to be heard in any judicial and administrative proceedings affecting the

[2] As of May 13, 2022, there are 196 "state parties" (countries) that have both signed and ratified the Convention. The United States of America has signed, but as of this writing, remains the only country in the world not to have ratified it.

child, either directly, or through a representative or an appropriate body, in a manner consistent with the procedural rules of national law. (*UN General Assembly*, 1989)

Based on this UN article and influenced by Hart's Ladder of Participation proposed for and published by UNICEF, Shier proposed a "Pathway to Participation" that comprises of five levels of participation:

1. Children are listened to.
2. Children are supported in expressing their views.
3. Children's views are taken into account.
4. Children are involved in decision-making processes.
5. Children share power and responsibility for decision-making (Shier, 110).

For these levels of engagement to be mutually beneficial to both parties, Shier states that three stages of commitment must be in place for each level: Openings, Opportunities, and Obligations. He explains that an *opening* occurs when a worker makes a personal commitment or statement of intent to work at each level in a certain way. It is only an opening, he adds, because at this stage, the opportunity may not be available (Shier, 110). The next stage—*opportunity*—is when the worker or organization is empowered to operate at any of the five levels in practice. These may include resources, skills and knowledge, development of new procedures, or new approaches to established tasks (Shier, 110). The final stage—*obligation*—takes place when the organization agrees on a policy that says parties should adopt the practice of each level. By working with youth at a specific level of participation, the collaboration becomes "built-in" to the system (Shier, 110).

### 5.2.1 *Children are Listened to*

Shier identifies this level as one when youth express a view, they are actually listened to, respectfully and with due consideration, by adults. He adds that this level requires that the youth initiate the discussion and that they are not solicited for an opinion. The stages of commitment apply to this level first by having the policymaker be ready and willing to listen; secondly, by requiring a structure that allows the discussion to take place such as a designated time and place without interruption; and finally, the third stage of commitment requires that listening to youth becomes a formal policy itself of the organization, thereby making it an obligation of the staff and decision-makers to listen to youth.

### 5.2.2 *Children are Supported in Expressing Their Views*

Unlike the previous level, this level seeks to elicit the opinions of youth. It recognizes that the viewpoints of youth are not readily offered since many youth feel either insecure, shy, or out of their leagues in expressing a view on a subject about which they feel they know relatively little. This commitment to positive action to involve youth and collaborate with them encourages them to express their views knowing they will be met with understanding and respect. The first stage of commitment for this level has the organization be ready, willing, and able to seek out the viewpoints of youth; the second stage of opportunity requires that the organization provide methods that are age-appropriate to eliciting opinions from youth. An interesting recommendation here is to use the "arts", specifically "creative visual methods". As we have seen in previous chapters and as we will see in upcoming chapters, youth prefer to express themselves with artistic tools (theatre, music, film, and digital art such as memes and filtered posts). Furthermore, to be inclusive at this stage, consideration for engaging with youth with disabilities or language barriers must be made. For the final stage of obligation, the organization must incorporate the methods of engagements recommended for stage two as part of official policy thereby committing the organization, its staff, and decision-makers to provide an environment in which youth are not only enabled, but supported and encouraged to express their views.

### 5.2.3 *Children's Views are Taken into Account*

This level begins with the affordances of the first two levels, but adds a component of accountability in which the policymaker actually takes into account the opinion of youth and has it directly influence and contribute to policy. This level of engagement ensures that tokenism does not come into play so that the appearance of engagement is made by listening to youth, but then never following up in a meaningful way with anything they suggest. This level adheres to Article 12.1 of the United Nations Convention on the Rights of the Child that clearly states that "the views of the child being given *due weight* in accordance with the age and maturity of the child". This does not obligate the policymaker to incorporate anything the youth propose into UN climate policy, but it does insist that whatever youth suggest is given full consideration. By being ready and able to listen to youth and to take their opinions into serious account, the organization satisfies the first two stages of commitment. The final stage of obligation is met with the organization making it a policy to implement Article 12 into its policymaking process.

### 5.2.4 *Children are Involved in Decision-Making Processes*

Perhaps most important in this enhanced model of youth participation in climate policy is this level that states that youth must be directly involved in policymaking processes. It indicates a transition from consultation to collaboration in policy creation. The previous levels recognize youth and invite contributions, but ultimately any actual decisions are still made by the policymaker. As Shier so eloquently puts it: "Decision-making remains the province of adults". In order to adhere to this level, some measure of power needs to be given to youth. While the UN Convention does not specifically mandate that youth be involved to this degree, Shier points out the many benefits that come with level of youth engagement "including the quality of service provision, increasing self-esteem, increasing empathy and responsibility, laying the groundwork for citizenship and democratic participation, and this helping to safeguard and strengthen democracy" (Shier, 114).

We can add to this list that when youth directly participate in the climate policy process, they are protecting their own futures with laws created by them and for them. The first stage of commitment is a relatively big one: organizations need to be prepared to provide an opening for youth to directly contribute to the policy process. The second stage requires an equally significant adjustment to the traditional way of creating policy. An opportunity must be made by the decision-maker to invite and work with youth which may require a major overhaul to the conventional structure of policy creation: times when they can meet (youth are often in school during regular business hours), language (consideration must be given to the vocabulary of youth which might be missing some of the language of policymaking), and venue (where do these meetings take place—in the policymakers' offices or in a more youth-friendly environment?) are just a few of the changes that must be made for this level to take place effectively. The final stage—obligation—comes when the organization makes it official policy to involve youth at this level and commits to revamping their infrastructure to accommodate this.

### 5.2.5 *Children Share Power and Responsibility for Decision-Making*

This is the highest level of youth engagement with the policy process and very similar to the previous level. It differs by having the adults in power willingly give away some of their power so that the interests of youth are not only invited and heard, but are included in policy. This provision is made at the outset; unlike the previous level that allows for participation in the policymaking process, level five mandates they share the power and take responsibility for the decision-making. In order for this to successfully take place, Shier recommends that the policymaker first consider all the risks and benefits of surrendering some of this power to youth and when the benefits outweigh the risks, then the organization moves ahead with making this happen in a "supportive environment" as described previously. This procedural innovation is often new to an organization and the process needs to be monitored to determine areas for improvement for future sessions. The first stage of implementing this level is for the organization to agree to share the decision-making power with youth. The second stage of opportunity exists when the organization creates a structure that allows for this to happen and invites youth to participate. The final stage occurs when the organization makes it a policy to share decision-making power and responsibility with youth.

Shier's model expands upon Hart's ladder and both identify the importance of respect, collaboration, and access. Shier goes a step further by using the UN Convention on the Rights of the Child's Article 12 to establish actual policy participation in the decision-making process. This can be seen as "authorship", an important component of youth participation as it directly reflects the needs and demands of youth and shares the responsibility of the creation of the policy. Youth become authors of their own climate policies and their representation in this manner is assured.

## 5.3 The Seven Ps

The final model upon which the Direct Approach is informed is known as the "Seven Ps" representing interconnected areas (referred to as "domains") of youth participation in the policy process—purpose, positioning, perspective, power relations, protection, place, and process—proposed by Helen Cahill and Babak Dadvand in 2018. The authors define this model as "a thinking tool for visioning, planning, enacting and

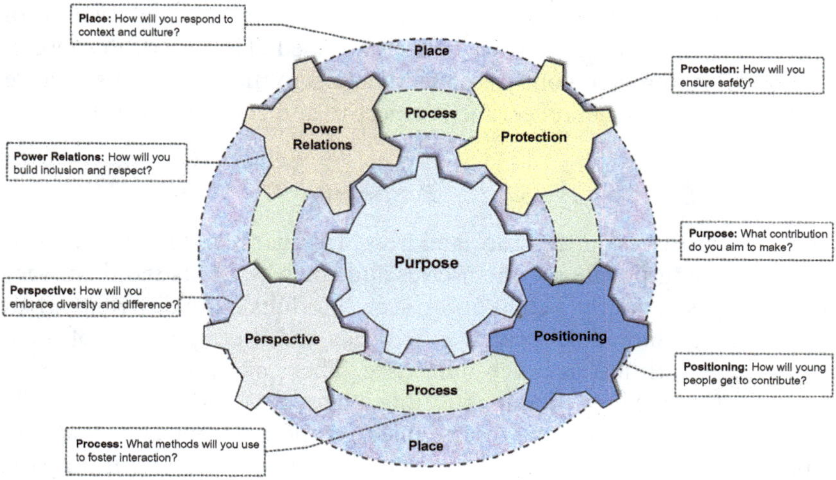


**Fig. 5.1** The 7P model: a framework for providing meaning youth participation in policy creation. (Cahill and Dadvand 2018)

evaluating youth participation" (Cahill, Dadvand, 248). This is the most comprehensive of the three including aspects of youth engagement with power not prescribed in the previous models (see Fig. 5.1).

### 5.3.1 *Purpose*

In this interconnective model, *Purpose* is the central cog in the wheel with all other cogs related to it in an interdependent manner. It is so positioned, according to Cahill and Dadvand, to suggest that Purpose is strongest when "collectively generated" and shared with young participants. "When young people are co-creators in the framing or in the re-shaping of purpose, they can share their ethics, vision and values, and contribute the motivating moral force which drives the program efforts" (Cahill, Dadvand, 248). This relates to the goal the global community of youth have expressed recently at COP conferences with respect to being consulted as a distinct society much the same way specific consultation is afforded to Indigenous communities and women's groups. Recognizing that youth have unique "ethics, visions, and values" acknowledges that policy created for them should also be created by them. Young people are not generally considered in the organizing structure of a policy creation session, but have been making in-roads with respect to being involved in discussions once the process has been established. The aim of Purpose in this model is to regard youth as equal to adults in the formation structure so that the process will reflect the perspective and culture of youth.

### 5.3.2 *Positioning*

The importance of youth culture is addressed in this second P—*Positioning*. The way young people see themselves and the way adults see them vary with the ages of youth: a child of 8 is seen by adults quite differently than a youth of 18 and adults perceive the needs and the experience of these ages groups with just as much difference. These narratives within youth culture determine how they are positioned by society and can prejudicially influence their involvement with policy creation. Chaill and Dadvand explain it this way: "…young people develop their sense of self, in part, through how they are positioned and treated within the social geographies of their everyday spaces. Therefore, the ways in which young people are positioned - either as leaders, advocates, investigators and co-contributors, or dependents, followers and passive recipients - not only conditions how they perceive themselves, but also affects how they develop a sense of agency and empowerment in relation to the wider social world" (Cahill, Dadvand, 249). It also conditions how adults, especially those in power, perceive them. Positioning, therefore, plays a crucial role in establishing the credibility of youth participants in policy discussions and their related respect and contributions from those in power who grant them access to this process.

### 5.3.3 *Perspectives*

This third P in the model reminds us that youth is not a common collective, but a tapestry of ethnic diversity, socio-economic profiles, social classes, educational backgrounds, and physical abilities. All these differences provide unique *Perspectives* of inclusion, participation, and contribution to the policy creation process and need to be considered in advance of giving "youth" a seat at the table. With this in mind, the authors tell us that "it is important to devise processes which actively address ways to reach, recruit and learn from the perspectives of those who might otherwise remain marginalised" (Cahill, Dadvand, 250). The UN recognizes

Indigenous communities as being separate and distinct cultures with their own traditions, beliefs, and practices and therefore acknowledge each contribution and perspective, not seeing Indigenous Peoples as a whole, but as a mosaic of cultures within a group. Youth need to be seen in a similar manner since the perspectives of youth in the Arctic community of Tuktoyaktuk would vary significantly from the youth living in Hong Kong, for example. However, both communities of youth are impacted by the common and global threat of climate change, so each should have their respective perspectives considered in any global climate policy.

### 5.3.4 *Power Relations*

The obvious relationship between those in power and young people is not the only *Power Relations* to consider when engaging youth in a policy program. As we see in school bullying, power dynamics exist among youth themselves. So too, do such imbalances exist within those government institutions tasked with policy creation. Who gets heard and when needs to be considered in advance and measures need to be established to ensure equal and fair representations among the prevalent and, at times, competing power relations in the room.

> It can help us acknowledge that not everyone has equal access to the discourses that allow them to take up certain positions. Some may adopt more dominant roles, whilst others may be relegated to sub-servient positions within the social space. This can be manifest in interpersonal relationships as well as within the institutional practices and structures. (Cahill, Dadvand, 250)

### 5.3.5 *Protection*

*Protection* is seen here as a consideration to be made when weighing the differences between "vulnerabilities and capabilities" of youth. In assessing the risk of involving youth in the policy process, it is important to determine if those in power are protecting youth participants to the degree that they may no longer be making a representative contribution. "Sometime a focus on the need for 'protection' can be used as a justification for limiting or overly governing the activities of young people, who are presumed either to be a danger to each other, or at risk from adults.

Whilst attention should be given to ensuring protective processes and measures, it is important to seek an appropriate balance between protection and participation…" (Cahill, Dadvand, 250). The Protection component recognizes the need to assess potential risks to youth participants and to install safeguards for them, but it also warns against taking no action against potential risks and excluding youth because of them.

### 5.3.6 *Place*

Where and how engagement between youth and power is the focus of *Place*. As mentioned previously, significant sectors of the global community of youth can be unintentionally excluded due to the fixed location of the policymaker, often in urban centres where many youth live, but not all. Financial restrictions impacting many youth as well as time commitments with school and summer jobs result in an inability to travel to these locations to physically participate. The recent pandemic has introduced a generally accepted form of virtual attendance through digital telecommunications systems like Zoom, Teams, and the pioneering platform, Skype, among others. This is a major contributor to resolving the issues of place and the related costs of travel; it still, however, does not address the potential digital divide that affects certain remote communities without access to WiFi or other digital technologies. Assuming these spaces can be occupied by youth, Cahill and Dadvand point out that there could still exist issues that may adversely affect meaningful participation related to the place of the policymaker: "Given that young people occupy physical, virtual and social spaces, it is important to think about how social and institutional settings can influence the possibility of participation. At a more physical level, a thinking through the domain of place may lead programmers to consult with young people about where they might gather, how they might gain access, and what relational, material or political 'terrains' need to be navigated in order to congregate and participate" (Cahill, Dadvand, 251). The possible intimidation factor of meeting in a government office—either physically or virtually—might tilt the power dynamic in favour of the policymaker. Meeting in a youth-friendly environment might have the same effect in the other direction. Ideally, therefore, a neutral place that is both functional and accessible should be considered when establishing a meeting place for both parties.

### 5.3.7 Process

The last P in this model is *Process*. This component considers the methods used to "invite and sustain" interactions between youth and policymaker. This becomes important in being fluent in the digital language that youth prefer to use in social engagements. Some of these platforms do not lend themselves well to formal meetings designed to forge environmental policy, yet some of them do. It becomes important to identify which methods are mutually beneficial and preferred at the outset of such sessions to ensure a working environment that is not foreign or intimidating to the youth participants, but are also functional and comfortable to the policymaker. Interestingly, the research of Cahill and Dadvand in this regard show great strides being made in methods that involve the arts: "Our own work has demonstrated the ways in which *drama-based* methods can assist young people to embody and thus express their visions for change, and to rehearse the strategies needed to advance from vision to action" (Cahill, Dadvand, 251). The example they provide is when participating youth are tasked with providing a proposed application of the policy outside the workshop and in everyday life. Here, instead of the traditional formal platform of business communications—the PowerPoint presentation—participating youth demonstrate how they would respond to the policy in their own communities through role-playing, a form of micro theatre presentation. "Rather than using PowerPoint presentations or flip charts, which would not be replicated out in the streets, role-plays were used to discuss and 'rehearse for life' and practice the micro interactions that were an important part of everyday life. In this, the processes used played a role in positioning, protection, power relations and perspective sharing" (Cahill, Dadvand, 251–252). The penchant for youth to express themselves to power through the arts is a recurring theme, as we have seen, in theatre (staged events and flash mobs at protests), music (concerts featuring songs and performers themed by causes), and films (documentary shorts produced by youth and screened to power to provide visible evidence that the written word and the spoken word cannot provide). These creative forms of expression prove to be comfortable safe zones of expression to youth when communicating directly with policymakers.

The 7 Ps is a valuable model of youth participation with power as it incorporates the strengths of the previous models proposed by Hart and Shier, and fleshes out some of the considerations suggested, but not directly addressed by the previous models such as Place and Power Relations. Collectively, all three models serve to inform a new proposed framework made in this book: *The Direct Approach*.

## 5.4 The Direct Approach

The three models examined in this chapter serve to inform a proposed framework which will be examined in practice through various new and emerging youth programs and opportunities for participation currently in place and being developed by the United Nations Framework Convention on Climate Change for its annual climate summits, the COP conferences. The second half of this book will examine how the *Direct Approach* dispenses with the angry voices of the protest marches and the digital demands in social media and attempts to establish a collaborative methodology that informs policy and engages youth directly with power. This is not yet the full community consultation afforded to other groups, but it is a step in the same direction as policymakers and youth climate activists work together to hear, inform, and influence each other on a global scale through developments in United Nations youth groups and engagement programs.

The previous chapter examined the ways in which youth are given "a seat at the table" in climate change policy creation. We looked at the ways the United Nations have tried to make the global community of youth a part of the annual climate summits, the COP conferences, through its YOUNGO constituency. We see how the UN and other national governments are attempting to listen to youth and now we will examine how—and if—youth are listening to them.

The proposed framework of the Direct Approach is as follows:

- *Access*: Members of the global community of youth need to know that invitations and opportunities exist for direct engagement with policymakers. Access to these opportunities must be provided by the policymaker through channels familiar to youth for meaningful discourse;
- *Respect*: Both parties need to enter policy collaborations with an open mind since neither party is fully aware of each other's perspectives, experiences, and cultures. Efforts to learn and understand these defining characteristics will facilitate discussions. Regarding youth as a separate and distinct global community, as well as its

inherent sub-cultures and differences within itself, acknowledge a culture that the policymaker needs to consider for fully inclusive and representative climate policy. Understanding the same sub-cultures of the policymaking community helps youth empathize with the government officials and can lead to solutions to their obstacles faster;
- *Collaboration*: Youth must be prepared to conduct research and present reliably researched evidence that may be absent from the policymaker's understanding of the community's unique relationship with the impacts of climate change. The policymaker, in turn, needs to be open to assigning such knowledge collection and accepting it as reliable representations of that community's culture;
- *Finance*: Funding programs and relative mechanisms need to be provided to supplement opportunities for direct engagement. Many young people cannot afford the travel to the policymaker's offices and many policymakers do not have a specific budget to travel to various and remote communities of youth leadership. It therefore becomes apparent that funding programs, especially at the UN level, need to be established to facilitate mutually agreeable places for engagement;
- *Authorship*: Youth need to be able to contribute directly to policy writing and feel they have some acceptable degree of ownership with the drafted policy. Review and approval phases for youth would contribute to this. This component ensures that the voice of youth is not only heard, but is accurately represented in any ensuing policy. Participating youth need to have their names attached to the authorship of any published policy.

This framework offers a supplemental structure to the Groundswell Approach's PLAN framework. With the ubiquitous availability of digital communications in most of the world, youth taking the Groundswell Approach are clearly achieving the goals of raising awareness and applying political pressure more effectively than ever before. However, without the Direct Approach being used as well, governments are reluctant to enact legislation which may adversely impact on the corporate sector that comprise most of their donor and electoral support. The Direct Approach augments the Groundswell Approach in a less adversarial way allowing for a participatory mode of representation of a politically marginalized community following the model of engagement in policy creation that is commonly afforded other communities. By collaborating rather than demanding, policymakers are more willing to act in a process that may take longer than conventional policy negotiations, but are more likely to begin than they would without it. In Chap. 4, we examined how the Prime Minister's Youth Council in Canada incorporates all five components of this approach to establish a successful strategy of engagement between youth and policymaker. What follows is a breakdown of the five components and how they can be used to accelerate the creation of progressive climate policy worldwide.

### 5.4.1 Access

We have seen on a national level certain programs and councils established specifically for youth climate engagement (Canada, South Africa, The Netherlands) and on an international level through the United Nations with YOUNGO, COY, and Action for Climate Empowerment (ACE). These examples serve as models for youth access to government policy participation, but, as noted, they are not without their faults. These problematic issues, however, serve as areas of improvement and provide opportunities to enhance the relatively new process structure between youth and policymaker. One of the greatest obstacles facing meaningful progress with access in place is the sincerity of the policymaker to engage in good faith.

When access is granted, it must be seized by youth to make government officials accountable for the promises they make to engage with youth. The tendency not to prioritize those too young to vote for politicians' re-elections must be recognized and countered with organized efforts to formalize their engagement with youth. The focus of securing that second meeting should come with an agenda of that second meeting, specifically, to address existing climate policy and to propose future climate policy. Ultimately, this is the goal of participatory politics for youth and by positioning themselves as a community that needs to be consulted before new policy is enacted, the second meeting becomes more focussed and makes best use of the access provided, even if it was initially granted as a form of greenwashing or tokenism.

### 5.4.2 Respect

It may seem natural to expect respect in any meeting, but with the political prejudices previously identified, it can sometimes be a challenge for young people to be taken seriously. Similar profiling by youth seeing politicians as untrustworthy, disinterested, and dismissive must also be checked so that both parties can actually work together productively, presumably the goal of such formal engagements. This process often comes with an early learning curve in which both policymaker and youth activist need to understand more about each other. The politician needs to understand the community of youth and its specific culture to fully comprehend its demands; the young activists need to understand the political obstacles and minefields that may be preventing the government from enacting the desired legislation. In many cases, the lack of understanding that prevents new climate policy comes from the government in terms of science. They may be able to read the "bottom line" of projected futures, but fail to grasp the data that led to these conclusions. This is a result of a lack of formal scientific training among many politicians.

There has been much scholarship on the comprehension gap between those who produce climate science data and those who produce climate science policy (Gavrieli et al. 2009; Stoutenborough et al. 2015; Meah 2019; Convery 2020). Most point out that politicians are not trained scientists and often lack the academic background to fully understand peer-reviewed scientific papers. The scientific community, as well, are responsible in part for their failure to find ways of successfully communicating data to the uninitiated without sacrificing accuracy. In their efforts to be precise and scientifically irrefutable, they end up providing "too much information"—much of it technical—leaving the policymaker overwhelmed with unfamiliar and incomprehensible material. Youth often experience a similar disconnect with climate science as well as policy creation and related politics. As a result, a mutual understanding of these shortcomings should lead to a mutual respect of both parties in the collective pursuit to craft progressive climate policy. The scientific community has a role to play here in demystifying the data for these two parties. As we will see in the next chapter, there are ways currently in use that are effectively helping to bridge the communications gap between science and policy for both youth and government officials.

A mutual respect of all parties, coupled with a sincere desire to improve or create substantive climate policy, will result in finding ways around communications gaps and other obstacles that have traditionally stalled significant climate action. The Direct Approach makes this affordance in a spirit of collaboration more than the Groundswell Approach does with its demanding rhetoric.

### 5.4.3 Collaboration

An essential element of the Direct Approach, both youth climate activist and policymaker clearly need to work together in the manner previously prescribed in order to yield positive results. But the nature of the said collaboration needs to be defined. In the case of the policymaker, they need to explain to youth the obstacles they face so that both parties may be able to find ways to overcome them. One of the common themes that arise when examining the relationship between the UNFCCC and youth is information gathering or "data collection". With the aforementioned communications gap between UN policymakers and the global scientific community serving as a consistent obstacle to climate action, the UN proposed a way of bridging that gap through a youth program embracing the philosophy of the citizen journalist: the *Youth Climate Report*. In 2010, Nick Nuttall, communications head of the United Nations Environment Programme at the time, asked for greater youth involvement in reporting climate research in the field around the world (Terry, 194). This project collaborates directly with all three principle players: the policymaker, the climate scientist, and the youth activist. This project yielded a collaborative relationship wherein the UN would suggest climate research themes (rising sea levels, greenhouse gases, droughts and flooding, etc.) each year and the global community of youth would respond with video reports on these topics that would feature interviews with climate researchers working in these fields. These short documentary reports assist the policymaker in bridging the communication gap between climate science and climate policy in two impactful ways: first, the visual medium of film provides the visible evidence of climate impacts and multiple video reports provide a global view of each issue; second, scientists speak with the young reporters of YCR differently than they speak with their peers. They address them in a more teacher-student dynamic which helps both youth and policymaker have a better understanding of complex scientific research. The success of this collaborative experiment in its early years has led to an evolution of technology and participation which will be examined more fully in Chap. 7.

In terms of collaboration, youth is seen to play an essential role not only in representing their global community to ensure their needs are being acknowledged and met, but also as reliable media representatives between science and policy, effectively serving both parties in productive ways that result in serving the aims and objectives of their own community in addressing climate change worldwide.

### 5.4.4 Finance

Another key element essential to success for the Direct Approach is finance. The ability for youth to physically participate in face-to-face meetings and actually occupy their seat at the table is often thwarted by a lack of funds for travel and accommodation at the offices of government or at international policymaking meetings like the UNFCCC's COP conferences. Young people are still in school for the most part and are either too young or not available to work full-time leaving them with little to no funds for independent travel and housing. While telecommunications platforms like Zoom, Google Meet, Microsoft Teams, and Skype have become popular ways of bringing people together online during the COVID-19 pandemic, they do not entirely replace the need to be in the same room for meetings that take place over a period of weeks. It has been noted that youth do not experience the same "participatory parity" as other non-state actors at the UNFCCC's climate summits, despite having their own UN-recognized constituency, YOUNGO (Thew, 2018) and this is largely attributed to a lack of financing.

> (There is a) lack of finances (that) prevents youth from capitalising upon available opportunities (*Access*) e.g. as volunteers self-funding their participation, youth struggle to apply for and prepare side events and exhibits in advance. This also restricts their ability to develop relationships with more powerful actors over time as self-financing repeat attendance often isn't possible. Youth attribute their participatory challenges to lack of recognition from other stakeholders despite being officially recognised as a constituency (Thew 2018). This further supports the theory that informal, invisible rules act as a barrier to participatory parity even when formal structures appear fair.... (Thew et al. 2020)

The lack of recognition identified by Thew et al. harkens back to the need for the *Respect* component of the proposed Direct Approach framework. While other non-state actors have access to funds to attend policy sessions such as the COP conferences repeatedly, lack of recognition often renders youth activists ineligible for necessary funding that may exist to participate in such sessions. Often, youth participation in UN climate events is funded privately through NGOs or Internet funding campaigns like GoFundMe. Sometimes, they are funded by governments home to the youth participants. In 2019, the federal government of Canada financially supported the travel and accommodation of six Inuit youth from

Tuktoyaktuk in Canada's Northwest Territories to attend COP25 in Madrid, Spain to screen a documentary short they made showing the devastating impacts of climate change on their community to delegates and media attending the conference (Terry, 198). This funding supports the *Access* component of the framework of the Direct Approach. Without financial mechanisms in place, access may be granted, but not always achieved.

A general lack of funds available for travel and accommodation to facilitate meaningful policy participation for youth is the purpose for which most funding related to climate change and youth is currently assigned. Many funding programs available to youth come in the form of awards in competitions or educational prizes for project creation. Canada's Youth Climate Change Fund might sound like a progressive government funding program to mobilize youth and encourage their active participation in policy discussions, but the available funds are only allocated for:

- Innovative academic research in the area of climate change for undergraduate/graduate work
- School/community projects on climate change
- Individual youth projects on climate change (Creech et al., 29)

Its more recent version, the Climate Action Awareness Fund, provides $206 million over five years to support Canadian-made projects that help to reduce Canada's greenhouse gas emissions, but involves youth only for programs that emphasize raising awareness, not policy participation (*Climate Action Awareness Fund*, 2022).

The Nigerian Youth Climate Coalition (NYCC), established in 2014, has mobilized more than 10,000 youth in Nigeria to participate in a variety of climate-related activities including international policy meetings at the UN's annual climate summits. While they are given access and clearly have the enthusiasm of a large community of motivated youth, their participation is limited due to a lack of financing. "Operating primarily as a team of volunteers, one of the NYCC's major obstacles is funding. Funding contributions have been received both as individual donations and project grant funding, yet the lack of consistent funding hampers the ability of NYCC to maintain some of its activities" (Benkenstein et al., 22).

Government funding programs intended to facilitate climate change policy participation, including those of the United Nations, are few and far between. Most of them exist outside of government in non-governmental organizations (NGOs), post-secondary educational institutions, and through the benevolent generosity of like-minded philanthropists. How, therefore, can governments and legislators be convinced to allocate the funding necessary to support the global community of youth in policy creation sessions that will ultimately affect this significant sector of the population directly? One way is to look at the impact youth have on policy when they are able to participate. These metrics can serve as quantitative evidence of success in representing the underrepresented. Further reporting can reveal qualitative evidence of success when considering the contribution the youth make to the content of the policy created. Most of the need for funds identified by youth are for travel to and accommodation at the places where policymakers meet and this represents a small amount of money relative to other funds that exist for climate science research and other related environmental programs. Academic institutions, particularly universities with strong research departments, already support conference travel and accommodations, and while many of their students are youth, this sector is still exclusionary of the vast majority of other youth who do not attend post-secondary institutions. Scholars like Harriet Thew and her vast body of work in youth climate activism and UN policy dynamics, echoes this proposed framework:

> ...there is a need for adult institutions to lend support (to youth climate policy participation). For example, funding could be provided and structures put in place to build the capacity of youth participants to engage with or secure the attendance of marginalised peers in their own countries and overseas. While YOUNGO does receive a small amount of designated funding for Global South participants, this needs to be substantially increased. Platforms with mechanisms for representation and accountability at local to national levels could also be devised, and regional meetings could be held where youth participants could foreground their local identities, knowledge, and experiences. (Thew et al., 2022)

The concept of foregrounding the position and experiences of youth communities on a national level is a productive mode of preparation especially when considered how it may inform UN policymakers prior to face-to-face encounters at multinational climate summits like the COP conferences. It also reduces the relative need for funding when gatherings to establish these declarations can take place at home. This is relative, of course, to the distance needed to travel within one's own country, but if the COVID-19 pandemic introduced anything to the way we engage with each other, it is that virtual meetings online are widely accessible and relatively affordable. Funding for the global UN climate summits still is required for physical attendance, but with the preparation of presentations proposed by Thew et al., a more productive and representative meeting of youth climate activists and potential policy partners exists.

### *5.4.5 Authorship*

The final component in the proposed framework of the Direct Approach policy is ownership through authorship. As well as attending a meeting with policymakers and advising them of your group's role to play in a specific climate policy, it is just as essential that the ensuing text of that policy be co-authored by youth climate individuals and delegations. This ensures that collaboration in the writing of the policy has taken place and that reviewed approval of the text has been granted by the youth participants. Similar to the *Participatory Mode* (Nichols 2017) or community filmmaking approach to documentary filmmaking, the profiled community contributes directly in the production of the film as well as the post production (editing, scoring, mixing). This process ensures that the story and representation of the community is accurate. Often the filmmaker is from outside the community and may unintentionally misinterpret the community's customs, traditions, and culture. The Participatory Mode safeguards against any such misrepresentation. Signing your name to the resolution or policy created in collaboration with the government official makes both parties accountable to its content and protects the community of youth from having their interests misrepresented or not represented at all.

A model of this methodology exists in the creation of the *Global Youth Statement* issued by YOUNGO at the COY conference that precedes the COP conference each year. The 2021 document released at COY16 was co-authored by 15 youth from 14 countries and edited by seven from seven different countries. The document divides its contributors into two sections: COY16 Policy and YOUNGO Policy, both representative agencies of the United Nations and the global community of youth. Authors of the 2021 *Global Youth Statement* for COY16 are: Aislinn Mulligan—United Kingdom; Irene Perez Beltran—Spain; Sophie Stroehler—Germany; Ahmad Bassam—Jordan; Shawn Jonczyk—Germany; Josephine Zimba—Malawi; and Almendra Caceres—Peru. Representing the YOUNGO team are Xuan Zi Han—China/Singapore; Kelo

Uchendu—Nigeria; Ester Galende Sanchez—Spain; Jan Kairel Guillermo—Philippines; Heeta Lakhani—India; Marie-Claire Graf—Switzerland; Abdallah Emad—Egypt; and Saher Baig—Pakistan. The editors, credited as "Final Readers", are Lana Weidgenant—USA/Brazil; Mark Terry—Canada; John Leo Algo—Philippines; Anh Tuan—Vietnam; Jin Tanaka—Japan; Xan Northcott and Serena Bashal—United Kingdom. This representation of youth contribution[3] presents the voice and perspective of youth to UNFCCC delegates and the world in matters of global climate change concern. A similar document without this authorship would only report on what is believed to be this community's stand with the possibility of under-reporting, misinterpretation, and misquoting or quoting out of context.

In addition to representational accuracy of the global community of youth, climate science and other content are refined with youth co-authors based on their own research and experiences with it. The collaboration with youth regards them as "active subjects" instead of "passive objects" and assists in the academic rigour of the process: "Involving (youth) participants as co-authors means that where there is disagreement, there must be recognition that it exists and resolution must be reached..." (Dunlop et al., 63).

But not all policy participation warrants youth co-authorship. A contributor status may be assigned when the actual policy uses the input of youth, but not their direct authorship of policy text. For example, when youth practice citizen science data collection with the supervision of climate scientists, the research they bring to an academic paper is credited to them and when incorporated into policy is attributed to them. This represents a form of authorship that is valuable, but not as clearly representational of the youth community. Another different form of authorship comes from youth who practise citizen journalism. Reports conducted by them and published by others will acknowledge their work through article bylines, photo credits, and video production credits. They may not have their names on the policy that uses this material, but they will be cited by name within the policy as contributors to the research and data upon which the policy is based. Examples of this can be seen from the entries in the annual Young Reporters for the Environment competition and the documentary shorts created by the UNFCCC's Youth Climate Report.

[3] Disclaimer: One of the editors of the *Global Youth Statement*—Mark Terry (author of this book)—is not a youth.

## 5.5 Conclusion

The Direct Approach benefits from the three previous models of youth engagement in policy decision-making that affects them. It also accelerates the path to progressive new climate policy when activists reach out to the policymaker—and when those in power reach out to youth—to establish these collaborative ecosystems. We saw in previous chapters how various governments are experimenting with this approach, most notably Canada with its Prime Minister's Youth Council. The next chapter will show how this works through an examination of other case studies where the Direct Approach is applied by other government and non-governmental programs.

## Works Cited

"About eBird", *eBird.org*. Accessed: May 11, 2022. Link: https://ebird.org

"Climate Action Awareness Fund", *Canada.ca*. Accessed: May 7, 2022. Link: https://www.canada.ca/en/services/environment/weather/climatechange/funding-programs/climate-action-awareness-fund.html

"Global Youth Statement", *United Nations Climate Change/YOUNGO*, Glasgow: 2021. Accessed: December 10, 2021. Link: https://ukcoy16.org/wp-content/uploads/2021/10/Global-Youth-Statement.pdf

UN General Assembly, *Convention on the Rights of the Child*, November 20, 1989, United Nations, Treaty Series, vol. 1577. Accessed: May 13, 2022. Link: https://www.refworld.org/docid/3ae6b38f0.html

Arnstein, Sherry R. "A Ladder of Citizen Participation." *Journal of the American Institute of Planners* 35.4 (1969): 216-224.

Benkenstein, A., Chevallier, R., Kosciulek, D., Lebea, D., & Worth, K. *Youth Climate Advocacy*. South African Institute of International Affairs, 2020.

Cahill, Helen, and Babak Dadvand. "Re-conceptualising Youth Participation: A Framework to Inform Action." *Children and Youth Services Review* 95 (2018): 243–253.

Convery, Frank J., and Gernot Wagner. "Reflections–managing uncertain climates: some guidance for policy makers and researchers." *Review of Environmental Economics and Policy* (2020).

Creech, H., Buckler, C., Innes, L., & Larochelle, S. (1999). A youth strategy for public outreach on climate change. *International Institute for Sustainable Development Business Trust, Manitoba.*

Gavrieli, Yael, M. K. Begin, Dov Hanin, Miriam Hern, Danny Levitan, Ran Levy, and Carmel Varnia. *Environmental Scientists and Environmental Policy Makers: Discourse Assessment and Action Recommendations* (2009).

Hart, Roger A. "Children's Participation: From Tokenism to Citizenship", *Innocenti Essays, No. 4.* Florence, Italy: UNICEF, 1992.

James W. Stoutenborough, Robert Nicholas Fette, Arnold Vedlitz, Carol L. Goldsmith, "Understanding the Communication of Climate Change Risk: Climate Scientists' Perspectives of Media Sources and Policy Makers", *Risk, Hazards, & Crisis in Public Policy*, March 29, 2015. Link: https://doi.org/10.1002/rhc3.12066

Lynda Dunlop, Elizabeth Rushton, Lucy Atkinson, Eef Cornelissen, Jelle De Schrijver, Tetiana Stadnyk, Joshua Stubbs, Chrissy Su, Maria Turkenburg-van Diepen, Fernanda Veneu, Celena Blake, Saul Calvert, Clémentine Dècle, Kirndeep Dhassi, Rosalind Edwards, Greta Malaj, Jovana Mirjanić, William Saunders, Yara Sinkovec, Suzan Vellekoop & Xinyue Yuan (2022) "Youth Co-authorship as Public Engagement with Geoengineering", *International Journal of Science Education*, Part B, 12:1, 60–74, https://doi.org/10.1080/21548455.2022.2027043.

Meah, Nafees. "Climate uncertainty and policymaking—what do policymakers want to know?" *Regional Environmental Change* 19.6 (2019): 1611–1621.

Nichols, Bill. *Introduction to Documentary*, 3rd ed. Bloomington, IN: University of Indiana Press, 2017.

Shier, Harry. "Pathways to Participation: Openings, Opportunities and Obligations." *Children & Society* 15.2 (2001): 107–117.

Thew, Harriet. "Youth Participation and Agency in the United Nations Framework Convention on Climate Change." *International Environmental Agreements: Politics, Law and Economics* 18.3 (2018): 369–389.

Thew, Harriet, Lucie Middlemiss, and Jouni Paavola. "You Need a Month's Holiday Just to Get over It! Exploring Young People's Lived Experiences of the UN Climate Change Negotiations." *Sustainability* 14.7 (2022)

Thew, Harriet, Lucie Middlemiss, and Jouni Paavola. "'Youth is not a political position': Exploring Justice Claims-making in the UN Climate Change Negotiations." *Global Environmental Change*, 61 (2020).

Terry, Mark. *The Geo-Doc: Geomedia, Documentary Film, and Social Change.* London: Palgrave Macmillan, 2020.

CHAPTER 6

# The Emerging Role of Youth at UN Climate Summits

The various models examined in previous chapters for youth participation in the policy process, including the Direct Approach, represent foundational frameworks for when youth actually do engage with those in power; however, how keen are policymakers to invite youth to the table for such discussions? What obstacles, real or perceived, are preventing organizations like the United Nations to implement these protocols and work alongside youth to create climate policies that they will inherit?

As we examined previously, climate change has become the cause célèbre of Millennials worldwide fuelled by their own social media platforms and fanned by extensive mainstream media coverage. Their well-documented protests and attendance numbers are impressive, but how effective are they in yielding the immediate and sweeping changes they demand. In his paper, *Fridays for Future's Disruptive Potential: An Inconvenient Youth Between Moderate and Radical Ideas*, Jens Marquardt casts doubt on their efficacy: "Although FFF has sparked debates about the urgency of tackling climate change, the movement's broader societal and political implications are yet to be seen" (Marquardt, 2). He suggests that the lack of results is related to a perceived threat by those in power that they may be replaced by the protestors and the radical ideologies they represent:

> A key motivator for social movements in general, and FFF activists in particular, is their high confidence in a future that is different from the estab-



M. Terry, *Speaking Youth to Power*, Palgrave Studies in Media and Environmental Communication,
https://doi.org/10.1007/978-3-031-14298-7_6

> lished routines and the business-as-usual scenario. These envisioned futures can be perceived as threats to existing, dominant power relations and incumbent interests. (Marquardt, 4)

Government officials, policymakers, corporate leaders, and others in power see young climate protestors more as an angry mob intent on dethroning them, rather than a group of concerned citizens simply asking for change. This perceived threat paralyzes power and is often one of the reasons why little to no progress seems to result in significant climate change policy creation. It becomes apparent that a different approach is needed to yield the desired results, an approach that takes into consideration this perception and is able to assure those in power that the threat to cancel them is not intended; only the desire to add progressive measures that benefit us all. Interestingly, there is research that suggests that there are those in power who welcome direct discussion with a goal of collaborating. A recent study of climate protests among young people in Canada found that Members of Parliament (MPs) found a *Direct Approach* would be "more effective".

A survey of protestors attending the Fridays for Future rallies in 13 cities found that of the largest age group, teenagers, only 10% had ever contacted a government official…"Each gathering features a large captive audience, who, though engaged in protesting, could be maximizing their impact by also writing short, hand-written letters to their MPs or leaving a phone message at their MP's office. Based on our experimental results and the interviews conducted with MP staffers, these interventions are expected to be more effective… In addition to generating necessary pressure on politicians to act on climate change, organizing group communications of this sort would also better educate members of the public on how to communicate with their representatives, allowing for lasting civic engagement" (Wynes et al., 17).

## 6.1 UN Youth Engagement Recommendations

This sentiment is echoed on a global level by the United Nations well before the Fridays For Future protests began. In a 2013 report by UNICEF UK entitled "Climate Change: Children's Challenge", the number one recommendation of eight made is that "(Youth) should be involved in the development of the climate change policies that will affect their lives through inclusive consultation mechanisms" (Burgess, 3). The report goes on to say that youth "will inherit the outcomes of today's climate change decisions. It has no voice in the process. Adults, who will not necessarily have to live with the implications of their decisions, are making climate change policies without consulting or listening to (youth)" (Burgess, 9). This argument is made by many youth activists, proactive changemakers, and governments and is often heralded as a key objective of climate or environmental justice campaigns. But while both parties seem to embrace this idea, the reluctance to establish such a formal structure still seems to persist.

In the wake of the 2021 United Nations climate summit in Glasgow, Scotland, COP26, there was much discussion on restricted representation among youth who felt their presence inside the convention centre was simply symbolic, providing no opportunity to collaborate directly in policy discussions for the few that were invited to participate. Marie-Claire Graf, Focal Point for Global North for YOUNGO, the United Nations' official youth constituency, addressed this issue directly in a United Nations Climate Change report just weeks before the conference: "While we see growing interest in youth engagement in climate governance, we need to empower independent youth-led mechanisms, which are recognized and valued as such. Youth-washing and youth tokenism needs to be addressed" (United Nations Climate Change 2021).

Following the conference, *Inconvenient Questions International*, a podcast based in Singapore, interviewed several youth leaders asking for their assessment of COP26. They all echoed the sentiments of Laura Bannon, Project Manager of the UNFCCC's Youth Climate Report: "Rather than putting all the responsibility in the hands of youth organizations (at the COP conferences), it's more about integrating (youth) into the process of policymaking" (Bannon, 01:00). The United Nations Climate Change office and YOUNGO released a *Global Youth Statement* at COP26 in 2021 which underscored this issue throughout its demands calling for "inclusivity" and "engagement" in the policymaking process for youth. The authors and editors of this statement acknowledge that youth now have a seat at the table, but they stress that their presence must be more participatory and less symbolic (*Global Youth Statement*, 2021).

This expressed desire to work more closely with youth and to directly collaborate with them came from the United Nations as far back as 2005 with the establishment of the UN Climate Change Conference of Youth (COY). The first conference was held in Montreal and is defined as a conference intended "to empower youth and formally bring their voices to the UNFCCC processes to shape the intergovernmental climate change policies" (Conference of Youth, 2021). The four major components of the Conference of Youth are:

(Creating a) Policy Document;

1. Capacity Building;
2. Skill-building Workshops; and
3. Cultural Exchange (Conference of Youth, 2021).

And herein lies the problem as identified by critics of COY. The components do not engage the delegates of COY directly with the policymakers of COP. Rather, they tend to suggest COY participants need to learn (items 3 and 4) and network (item 2) and practise writing their own policies (item 1). None of these components address collaboration or meaningful policy representation with world leaders, their policymakers, or other United Nations officials who work on UNFCCC policy and communications.

At COP26 in 2021, many attendees of COY questioned why their conference needed to take place before the doors open to the COP conference. Many expressed this segregation as "exclusionary" and pointed out that the UNFCCC often hypocritically uses the term "inclusivity". COY representatives felt that this term was not being applied to them. Yolanda Wright, Global Director of Child Poverty and Climate at Save the Children, said: "For decades, (youth) have been routinely excluded from and overlooked in decision-making on the climate crisis. Despite promises that (COP26) would be the most inclusive ever, young activists have already described it as the complete opposite" (Press Release, 2021).

With the COY conference occurring before the COP conference instead of during it, many believe the UNFCCC is insincere in its claim for youth inclusivity giving rise to a new term to describe this: "youth-washing", coined by reporter Alleen Brown in an article she wrote for *The Intercept*, an online news organization "dedicated to holding the powerful accountable" (About the Intercept, 2022). Her article, *How the Fossil Fuel Industry is Attempting the Buy the Global Youth Movement* (Brown, 2019) was published December 13, 2019 just as COP25 concluded in Madrid. In her article, she reports on a meeting of CEOs from fossil fuel giants BP, Shell, and Equinor during the Oil and Gas Climate Initiative in New York. In a surprising move, they attended a lunchtime meeting with representatives of Canada's Student Energy that gave youth an opportunity "to grill the CEOs about their inaction on climate change" (Brown, 2019). She uses this meeting as an example of "youth-washing", an insincere gesture to suggest the fossil fuel industry is actually listening to youth climate activists, but in the end, ignoring them.

"With 'youth' becoming synonymous with climate action, corporations and politicians are increasingly using young people to portray themselves as climate serious" (Brown, 2019).

This may well be the case for some industries and even some governments, but the UN is endeavouring to find ways of actively engaging them with the policy process, short of having them in the Blue Zone at the COP conferences actually authoring policy beyond their own *Global Youth Statement* each year.

## 6.2 Canada's Youth Policy

Some countries are recognizing this oversight and are introducing programs and mandates that intend to not only afford youth opportunities to be heard, but to involve them directly in policy discussions and creation. In 2016, Canada established a Youth Secretariat specifically to address the needs of one-third of that country's population. This office has issued a document describing its aims and objectives called *Canada's Youth Policy.* The guiding principles of the policy are:

- Youth have the right to be heard and respected;
- Youth have the right to equal access to opportunities and supports;
- When youth reach their full potential, it benefits all Canadians (*Canada's Youth Policy*, 2022).

Specifically addressing the United Nations Convention on the Rights of the Child's (UNCRC) mandate that youth "have the right to an opinion and adults have a responsibility to listen and take them seriously" (*Canada's Youth Policy*, 2022), the policy acknowledges the equality and diversity challenges that have been identified as barriers to meaningful youth engagement in policy participation: "sexual orientation, gender identity, race, ethnicity, religion, disability, education, income, culture, geography and age" (*Canada's Youth Policy*, 2022). As there are many issues related to youth, Canada's Youth Policy identifies six areas of "priority":

- Leadership and Impact;
- Health and Wellness;
- Innovation, Skills, and Learning;
- Employment; Truth and Reconciliation; and
- Environment and Climate Action (*Canada's Youth Policy*, 2022).

Most significantly with respect to this book's investigation, the area of *Environment* and *Climate Action* is worded to suggest that policy participation is a component of this priority and not just climate change information exchange. The mandate of this focus is a commitment that considers the key factors of what this book defines as a framework for the Direct Approach—a structure that addresses the barriers and establishes the considerations necessary for incorporating the global community of youth into meaningful climate policy participation:

> Environment and climate change is a top priority for many young Canadians who are advocating for increased protection and conservation of the environment. Youth are conscious of the negative impacts climate change has on individual and community health, and recognize that they will be the generation to deal with the consequences. Youth are eager to have more of a say in environmental decisions and policies in order to create a Canada that reflects their environmental values. They want a Canada that protects its natural environment and addresses climate change in a process that emphasizes reconciliation with Indigenous Peoples. Young Canadians are motivated to find innovative solutions to environmental challenges, promote sustainable practices and lifestyles, and move towards a green economy in ways that respect the rights and values of Indigenous Peoples in Canada. Youth want to see further immediate action to protect the environment so that they, and future generations of Canadians, can inherit a healthy world (*Canada's Youth Policy*, 2022).

This policy recognizes the importance of participation in policy creation youth calling it "a say in environmental decisions and policies" and addressing the community's unique perspectives and expectations by acknowledging policy that would create a country "that reflects *their* environmental values". In citing the UNCRC and by declaring that the voices of youth need not only to be heard but to be "respected", the Canadian government acknowledges one of the main barriers identified in this book and by youth climate groups around the world that their age and inexperience do not make them credible participants in the policy process.

## 6.3 Canada's Youth Climate Lab

This might seem like a progressive ideology for youth climate activists on paper, but does it get put into practice, and if it does, how successful is the action? Supporting this mandate is Canada's Youth Climate Lab (YCL), a non-profit organization founded in 2017 that aims to create innovative approaches to engaging youth directly in the climate policy process. Unlike most governmental organizations such as the United Nations and Canada's Youth Secretariat that do not provide funding to youth individuals and groups to facilitate their involvement in policy participation, the YCL does. In its first three years of operation, the YCL collected donations amounting to $450,000 (USD) and dispersed $120,000 (USD) of that "directly into participants' hands for their climate action" much of it covering travel expenses to COP conferences (*Our Impact*, 2022). With the financial support and partnership of the International Development Research Centre, a Canadian federal Crown corporation that funds research and innovation, the YCL and Ghana's Green Africa Youth Organization (GAYO), were able to work together to create and deliver succinct actionable policy recommendations at the 2021 UN climate summit held in Glasgow, COP26.

> Leading up to this global event, GAYO held six online summits bringing together over 200 youth activists from Asia, Africa and Latin America to discuss and recommend how to spur youth-led development and climate-change adaptation and mitigation. Regional and international qualitative reports synthesize their findings. In addition, audioblogs present interviews with 10 young climate leaders from the Global South, and a brief synthesizes their insights. The consultations informed the panel discussion featuring Gulugulu and four other young climate leaders as part of the Resilience Hub at COP26. (*Youth Claim*, 2022)

This participation would not have been possible without the financial support of the International Development Research Centre (IRDC) and this demonstrates that access, when granted, can best be achieved when funds are no longer a barrier to participation; a key component of the Direct Approach—*Financing*. The information provided at the panel showcased the work of youth as well as their position on current and proposed climate action. Led by two youth groups—one in Canada and one in Ghana—the virtual summits were sure to include 10 youth climate leaders from the Global South as well as representation from 200 other youth climate activists from three continents to provide a more inclusive global presentation to policymakers. This engagement with UN policymakers and other environmental changemakers at a UN climate summit serves as an excellent model for representing the global community of youth in consultation and collaboration of climate change policy discussions and creation. It also serves as an example of how to engage in participatory politics between youth and policymakers.

## 6.4 Sustainable Orillia

In addition to initiating and facilitating opportunities for youth engagement, organizations like the YCL also conduct research of the times when youth are invited by power to participate. One government program called *Sustainable Orillia*, an initiative of the City of Orillia in Canada that works with citizens to develop sustainability strategies ("Who We Are", 2022). This program affords youth the opportunity to participate in developing the city's sustainability policy.

The inaugural intergenerational policy session featured youth participants who were asked to present their research and make suggestions from their perspectives. Two of the youth participants were Blythe Wieclawek, a grade 11 student at Twin Lakes Secondary School in Orillia, Ontario, Canada and Weiqi Xu, a Grade 12 student at the same school and both members of Sustainable Orillia's Youth Council (SOYC).

Wieclawek was impressed by the size of the youth council when she first joined and after a couple of years, became its president.

> When I joined the council it was in its second year of running and what pulled me in was the size and energy of the group, being about 20 members. However, most of our initiatives at that time were awareness-based, as that was what most members were most comfortable with. When I became the president at the end of last year, I knew that I wanted to pull the council towards more advocacy work because that is what is proven to create real change. It did not take much convincing to get the other youth council members on board—young people are ready to work with and pressure local governments to enact climate policy. (Wieclawek, 2022)

Praising the city's desire to make its youth council truly autonomous and youth-led, Wieclawek recalls feeling empowered when assigned the role to participate directly in an initiative called the City of Orillia's *Climate Change Action Plan.*

> The city conducted a focus group with the youth council and asked us for our views on sustainability and our opinion on Orillia's proposed plan. The City was looking to create a plan that was as ambitious as possible while still remaining economically feasible. I specifically took on a larger role in this project as I was the youth representative for the City, so I was on all the planning calls that occurred to develop the three choices. I was able to work directly with city officials, and used my voice multiple times to influence the City to create the most ambitious climate change action plan possible. The City ended up creating a pretty ambitious plan for a small community—Net Zero corporate emissions by 2040 and Net Zero community emissions by 2050. Myself and the council definitely had to fight for the plan to be as ambitious as possible, but the City actually listened to us and gave us space to voice our concerns. This experience definitely empowered myself and my group—having adult organizations trust us to put input on large-scale projects such as these instills a lot of confidence. Youth Climate activists do this work to protect our futures, and having adults and City officials acknowledge our motivations and work with us to solve this problem is essential to create the scale of change needed. Youth have such bright ideas and so much energy, but it is hard for us to break into adult-centric spaces, such as government spaces. While youth climate activists will continue to advocate for our environment and our futures regardless of adult support—we must, because of everything that is at stake—having support from adults helps ease the burden and make our initiatives more successful, especially those that involve pressuring governmental organizations. It is not only one generation's job to solve the climate crisis—everyone must do their part, and that is why adult-youth collaboration is so important. (Wieclawek, 2022)

The city ended up creating a pretty ambitious plan for a small community—Net Zero corporate emissions by 2040 and Net Zero community emissions by 2050. Wieclawek recalls having to fight for the plan to be as ambitious as possible, "but the city actually listened to us and gave us space to voice our concerns". In the end, everyone in the youth council felt encouraged to continue fighting climate change through this Direct Approach.

> This experience definitely empowered myself and my group—having adult organizations trust us to put input on large-scale projects such as these instills a lot of confidence. Youth climate activists do this work to protect our futures, and having adults and City officials acknowledge our motivations and work with us to solve this problem is essential to create the scale of change needed. Youth have such bright ideas and so much energy, but it is

> hard for us to break into adult-centric spaces, such as government spaces. While youth climate activists will continue to advocate for our environment and our futures regardless of adult support—we must, because of everything that is at stake—having support from adults helps ease the burden and make our initiatives more successful, especially those that involve pressuring governmental organizations. It is not only one generation's job to solve the climate crisis—everyone must do their part, and that is why adult-youth collaboration is so important. (Wieclawek, 2022)

Her fellow council member Weiqi Xu recalls the frustration that youth had expressed in engaging in a meaningful way with those in power. He and others were not confident they would have the opportunity to contribute directly to the policy process in a non-tokenistic participatory manner. "Most of the climate conversations I have with other youth deal with things we disapprove of—like the Bradford bypass highway", said Xu. "We want people in charge to listen to us. We can discuss all we want about what's happening across the country, and we do – all young people know about this stuff – but it's really a matter of what we can do to help salvage our future" ("Local Youth", 2022).

He explained that the government's youth council encourages sustainability activism among its youth members through social media campaigns and helps facilitate direct, "seat-at-the-table" policy participation with policymakers. "What we do in the SOYC is advocate for climate action from youth-based perspectives through many different media be it social media, climate rallies, youth summits, or providing youth perspectives (directly) to the city and (its) council" (Xu, 2022). His experience with the city's Climate Change Action Plan echoes the experience of his fellow council member Wieclawek: "The city listened to us", he said. "I feel that our opinions were respected, and were taken into consideration during the process. The city stayed in touch throughout the process to update the SOYC" (Xu, 2022).

Xu feels his experience shows a genuine sincerity for those in power to engage youth in climate policies that will affect their futures and is encouraged by the process. "I feel like the fact that the city reached out to youth to get their personal opinions on the issue of climate change and what they would like to see in our city's climate change action plan made us all within the SOYC feel empowered. After that, we participated in a *phone zap* to several ministers and had a social media campaign to try and get the ministers attention to the Bradford Bypass project which we demanded be stopped or at least be reviewed properly" (Xu, 2022).

### *6.4.1 The Phone Zap*

A "phone zap" is another new telecommunications advancement that youth climate activists are using to directly connect with those in power. A quick and succinct form of digital protest, a phone zap—sometimes referred to as a "call-in campaign"—is an easy and direct way of letting public officials know what you want them to do or to stop doing. The steps in creating a phone zap are as follows:

- **Write a Concise Call to Action**. You are speaking most likely to supporters or allies who don't need your polemic or additional convincing. If there is a lot of additional background to convey, you can link or reference another article or post for those who need it. A giant rambling wall of text turns off motivation and engages next to no one.
- **Write a Sample Script**: Plenty of people are self-conscious and a bit reluctant to call authority figures at some hostile institution. Throwing out an example of a brief message conveys the talking points, reassures participants, and gives them something to follow or expand on.
- **Select Your Targets**: Once you've identified the targets, usually government officials related to your cause, call them and make sure the numbers are still active and correct. There's nothing worse than trying to build a campaign, sending out a thousand emails with wrong or dead numbers on them.
- **Create a Hashtag**: A hashtag will help spread the word and help yourself keep tabs on who's committing.
- **Provide Multiple Numbers**: Once someone commits to participate, they are often happy to call multiple numbers while they are at it. Rank them in order of importance and most importantly, test them. Also, when all sorts of people are calling and taking over these offices with calls, it is often hard to get through. Give people a selection of numbers to call and they will keep trying, jumping from one number to the next—more calls land, more voicemail boxes get filled, and more pressure is applied.

- **Promote the Zap**: Advance notice and constant promotion makes an easily lost request into an event that looks worthwhile and actually is strategic, thoughtful, and supported. As with any online outreach or social media campaign, consider when people are logged on. Use multiple channels and include the hashtag you picked on social media, email listservs, text messages, face-to-face meetings, and websites.
- **Set a Target Time Window**: Set a day to shut that office down. Calls trickling in over a week don't make an impression. Make that request for a supporter's time as concrete as possible. You can even have specific groups adopt chunks of time, ensuring you keep phones ringing throughout the day.
- **Provide live updates**: For the day of the zap, have one or more people commit to being on social media to post real-time updates. This adds to the zap's visibility and keeps momentum up for people calling in ("How to Organize a Phone Zap", 2018).

The example from Orillia reveals the positive outcome that youth take away from such participatory policy sessions. They not only make a direct contribution to a city's climate and sustainability mandate, but through the respectful and invited experience, feel empowered and encouraged to take on the "world". It also is a strong example of the Direct Approach in action: *Access* was given to youth; *Respect* for the research and opinions of youth is provided; *Collaboration* in the policy process took place when those in power relied on the research presented by youth; *Finance* was considered but not necessary since attending the meetings in a local place was easily accessible for all youth participants; and *Authorship* will be provided in the action plan acknowledging the participation of the youth roundtable.

## 6.5 The UK's Students Organizing for Sustainability

Another example of the achievements that are possible for youth in working directly with the policymaker when finances are less of a barrier, is the policy statement created by the campaign group Teach the Future (TTF), which is hosted by the UK-based Students Organizing for Sustainability (SOS UK). TTF is a youth-led campaign to repurpose the education system around the climate emergency and ecological crisis ("Policy

Documents", 2022). This initiative recruited six young climate activists between the ages of 18 and 25 and provided training in policy writing to familiarize them with the process. This part of the program aims to eliminate the objection often made to youth participation that points to their lack of policy experience.

Following their training the group drafted a comprehensive sustainability policy presenting it to Members of Parliament in England for review and refinement. Once the jointly drafted policy statement was created and was in the form of a bill called the *Climate Education Bill*, it was presented at the House of Commons giving the six young authors a unique opportunity rarely afforded most people in seeking approval of a bill at this level, let alone young people. The bill was not passed, but "several of its components were used in other policy applications" (Philippa Kos, 2022), and most significantly, it was a key influence in the Department for Education announcing their first Climate Education Strategy for England at the United Nations climate change conference, COP26, in Glasgow Scotland in 2021. At this conference, members of the TTF campaign had another rare opportunity to present their policy proposals at a series of meetings with delegates, negotiators, and international environmental policymakers.

## 6.6 Blagrave Trust: Supporting Young People to Be a Part of the Policymaking Process

Funding for this unique initiative came from the Blagrave Trust, a UK-based philanthropic organization, one of very few organizations that specifically funds projects for youth that support their participation in policy creation. The Blagrave Trust gave TTF a grant of £10,000 in 2020 to assist them in drafting their Education Bill, followed by a second grant of £25,000 in 2021 to support their wider, growing policy work including attendance and engagement at COP26.

The focus on policy education, creation, and participation at policy events such as presenting bills at the House of Commons and the UN climate summit, COP26, is a significant, yet rare, contributor to the Direct Approach's component of *Finance*.

Over the course of 2021, the Blagrave Trust team identified supporting young people to be a part of the policy-making process as one of three strategic funding areas; and not just for climate change policy, but all issues of importance to youth. They have since established a fund dedicated for this purpose, which was launched in early 2022.

While recognizing the importance of policy creation education and working alongside the policymaker in a meaningful manner, Blagrave also establishes an admirable standard for youth climate financial support by funding training programs for following up with policymakers to keep them accountable and for the previously identified need to cover travel expenses to the COP conferences. As a further example, they also provided a grant to *Restless Development* (a non-profit global agency that supports the "collective power of young leaders to create a better world") to support a partnership with the UK Youth Climate Coalition (a UK-based a non-profit youth organisation that mobilises and empowers young people to take positive action for global climate justice) and their presence at COP26. This grant was for £30,000 and supported training for the young COP attendees in accountability mechanisms and how to follow up with policymakers on what they heard and committed to at the conference; and associated costs for attendance and participation.

Financing initiatives such as this are instrumental in advancing the aims and goals of the Direct Approach by further facilitating the opportunities for youth to participate in policy discussions, collaboration, creation, and presentations with those in power. It also represents a step forward in the demands of the youth who participate in the Groundswell Approach for climate action.

## 6.7 The National Youth Council of Ireland

Another national mandate for youth involvement in climate policy exists in Ireland: the National Youth Council of Ireland. In their climate policy they state that youth engagement in policy creation is a "key issue" for the council because "young people have the right to participate in decisions shaping their present and future" and "it is young people who will be most impacted by how successful we are in meeting targets set out in the (Sustainable Development Goals), the Paris Climate Agreement, and other local, national and global agreements and strategies" (Duffy et al., 9).

In many ways, the Direct Approach can be achieved on relatively manageable municipal and national levels, but how does it work on an international level with a government organization as big and complex as the

United Nations? Clearly, there are more obstacles to overcome, but as we will see, the UN is committed to developing programs and strategies that address these barriers and make youth participation more inclusive and meaningful at its COP climate conferences.

## 6.8 The United Nations and Youth: YOUNGO, COY and ACE

With the possible exception of *Finance*, the United Nations is adhering to all of the components of the proposed framework of the Direct Approach for youth participation in global climate policy. They do this by making opportunities for engagement available at the COP conferences through the United Nations Framework Convention on Climate Change's three main youth agencies: YOUNGO, COY, and ACE.

### *6.8.1 YOUNGO*

YOUNGO, the official youth constituency of the UNFCCC, was created in 2009 and confirmed as an official constituency in 2011. The most current numbers show that YOUNGO has a member base of more than 200 youth NGOs, more than 5500 individuals, and more than 20 working groups. Membership defines youth as anyone up to the age of 35. There is no minimum age. The guiding principles of YOUNGO, last updated on August 2, 2021, follow the components of the Direct Approach with the possible exception of *Finance*. While YOUNGO does not fund travel and accommodation to COP conferences directly, for example, many of its members do and in the spirit of collaboration, those seeking such financing can submit a request to the digital listserv reaching all members in the database. Those who wish to support the request are free to do so. Here is the complete list of *YOUNGO's Principles*:

*Climate Action*
We act from passion, purpose, compassion, hope, and commitment to the cause. As so, we are action-oriented and value proactivity, always focusing on our shared goal of raising ambition to secure intergenerational equity, climate justice, and sustainable development.

*Openness*

We welcome every young person who wants to constructively contribute to the constituency. We aim to provide equal access to opportunities to all engaged people, so everyone has a chance to meaningfully engage with our activities and the international decision-making spaces. We also work in cooperation and collaboration internally amongst all our members and externally with relevant stakeholders.

*Inclusiveness*

We represent the voice of youth around the world, embracing cultural and social diversity in all of its aspects – in terms of age, gender, race, religion, ethnicity, ability, sexual orientation, gender identity, and others. We also call for equal rights and meaningful participation, especially encouraging historically marginalized groups and vulnerable communities to come to the forefront for self-expression and representation.

*Dignity*

We promote an empowering and enabling environment for youth, in which all people are recognized for their inherent humanity and treated ethically. Hence, we do not accept any kind of harassment – be it sexual, physical, or psychological – and hold ourselves and each other accountable through a zero-tolerance policy.

*Horizontality*

We consider all members as equals in their differences, having the right to speak and to be heard. No individual should hold control and power, develop entitlement, or claim authority of the collective. Yet, we have a governance structure, decision-making processes, and specific roles and mandates, that secure an open and constructive dialogue, and that shall be respected in order to have coordinated and effective impact.

*Transparency*

We promote open and honest communication, not keeping, concealing, or restricting information that is essential to the functioning of the constituency. Meanwhile, we also respect sensitive, confidential, or personal information of our members. This way we seek to have an open, safe, and fair environment for all to participate.

*Integrity*
We work in service of our collective, operating with responsibility and commitment. We shall always seek to engage with impartiality, selflessness, and loyalty to the constituency, avoiding conflict of interest and aiming towards the collective benefit (*YOUNGO's Principles*, 2021).

In reviewing the key components of the Direct Approach against the guiding principles of YOUNGO, it is evident that the UNFCCC intends to establish an infrastructure that involves youth in the policy process. Specifically, YOUNGO provides *Access* to UN climate summits and its global network of environmental policymakers under its principle of *Openness* whose aim is to "provide equal access to opportunities to all engaged people, so everyone has a chance to meaningfully engage with our activities and the international decision-making spaces". In adhering to the Direct Approach's call for *Respect*, YOUNGO's principles of *Inclusiveness*, *Dignity*, and *Horizontality* all call for this by "embracing cultural and social diversity in all of its aspects – in terms of age, gender, race, religion, ethnicity, ability, sexual orientation, gender identity, and others. We also call for equal rights and meaningful participation, especially encouraging historically marginalized groups and vulnerable communities". For the Direct Approach's requirement of *Collaboration*, YOUNGO's principle of *Openness* specifically addresses this by stating that it works "in cooperation and collaboration internally amongst all our members and externally with relevant stakeholders" ("YOUNGO's Principles", 2022). For the final component of the Direct Approach, *Authorship*, there is no guiding principle that specifically addresses this, yet we know they publish policy documents including the names of its youth contributors, authors, and editors.

Fundamental to the complaint of youth today within YOUNGO is that while they may have access, meaningful "participation" is still denied them. Member groups of YOUNGO "are bound by the rules of observer participation, able to watch but not participate in negotiations and granted the same participatory opportunities as other constituencies. However, their participatory strategies and the level of agency they possess is not yet understood..." (Thew, 372). Some accredited YOUNGO delegates to the COP conferences may have access to the convention centre, but not all of its meetings and events for reasons of political sensitivity for the most part. But when they are denied entry to high-level policy and negotiation meetings some feel like they are being "treated like a second-class citizen" (Thew, 377).

### 6.8.2 *Conference of Youth (COY)*

With the first Conference of Youth (COY) taking place in Montreal, Canada in 2005, the UNFCCC event is now the world's largest "youth conference related to the multilateral UN climate processes" (Conference of Youth (COY), 2021). While the conference does not directly provide access to participatory politics and direct collaboration with environmental policymakers—or even the COP conference for that matter—it does provide a valuable "space for capacity building and policy training", essential knowledge and preparation for when youth are provided access to policymakers and for when they contribute to their own policy paper, the *Global Youth Statement*. Members of YOUNGO and others have been critical lately of COY taking place *before* the COP conference and *during* it. They see this as hypocritical to the UNFCCC's claims of inclusivity.

COY conforms to the Direct Approach in terms of *Collaboration* by convening a conference for its members, but not directly with the policymakers. *Respect* is also evident in its mandate for conference attendees, but again, just with each other, not with the policymaker. *Finance* is addressed in a unique way. COY does not provide funding directly, but through YOUNGO membership support, funds may be secured through member NGO programs. As well, in order to limit the costs of international travel, COY has established Virtual COY (vCOY), Regional COY (RCOY), and Local COY (LCOY). Virtual COY "is the virtual space where the global youth community facing different barriers such as lack of travel funds, visa restrictions, age barriers or due to environmental considerations comes together" (Conference of Youth (COY), 2021).

Regional and Local COYs "localize the climate conversation and gather more inputs from the ground. RCOYs and LCOYs discuss and develop respective outputs, which feed into national and regional climate policy making as well into GCOY and consequently into the climate negotiations" (Conference of Youth (COY), 2021). Again, we see a lack of direct access to the policymaker, but COY uses its various meetings to "feed" the contributions of attending youth directly to national and regional climate policy. This, in turn, helps inform the global climate negotiations for each country to which COY members contribute policy collaboration.

While the Groundswell Approach has captured the passions of youth around the world and provided them with activities to voice their frustration with government inaction in large numbers, both in virtual and in-person protests, there are aspects of this unprecedented show of solidarity that is counter-productive. Some government officials fear the groundswell and feel threatened that their jobs may be on the line resulting in detachment from those who, at the moment, represent a group that are often too young to vote. Public opinion is certainly paying attention to the global community of youth as their numbers and voices are hard to ignore, but the actual goal of realizing progressive climate-related policy is still out of reach for many protestors.

On the other hand, the Direct Approach appears to be a mutually desired methodology embraced eagerly by youth and cautiously by policymakers. The United Nations have opened its doors to youth through YOUNGO and COY in this regard, but just a crack. The COP conferences remain forums of climate policy deliberation from which youth feel excluded. They may make presentations at the COP conferences, but more often in the Green Zone, a space reserved for youth groups, civil society, academia, artists, and businesses to host events, exhibitions, cultural performances, workshops, and other presentations. Taking the Groundswell Approach, large numbers of youth engage in COP activities in the Green Zone, but more along the lines of "awareness raising" than policy participation, collaboration, and consultation. Those activities are in the Blue Zone, a UN-managed space which hosts policy discussions and negotiations on national, sub-national, and multinational levels. This space brings together delegations from 197 Parties (countries), alongside observer organizations to propose resolutions, issue collective statements, and draft policy that is ratified by member parties. This is where youth want to be and small steps are being made to accommodate this.

As we will see in the next chapter, those in power and meeting in the privileged Blue Zone are working with youth by assigning themes and topics for youth to report on in the form of citizen journalists. The education and outreach objectives of Article 6 introduced in 1992 and Article 12 introduced in 2015 are flowing in two directions: from the top down and from the bottom up as youth and policymaker inform and consult each other to supply understanding and knowledge mobilization that accelerates the path to progressive climate policies at these UN climate summits.

Article 6 of the United Nations Framework Convention on Climate, first established in 1992, aims to address education, training, and public awareness of climate change. It does this by promoting, facilitating, and cooperating in the following:

(i) the development and implementation of educational and public awareness programmes on climate change and its effects;
(ii) public access to information on climate change and its effects;
(iii) public participation in addressing climate change and its effects and developing adequate responses…
(iv) training of scientific, technical, and managerial personnel;
(v) the development and exchange of educational and public awareness material on climate change and its effects; and
(vi) the development and implementation of education and training programmes, including the strengthening of national institutions and the exchange or secondment of personnel to train experts in this field, in particular for developing countries ("Article 6", 1992).

Related to this is Article 12 of the Paris Climate Agreement which reinforces the main principles of Article 6, but in a more condensed manner:

> Parties shall cooperate in taking measures, as appropriate, to enhance climate change education, training, public awareness, public participation and public access to information, recognizing the importance of these steps with respect to enhancing actions under this Agreement. ("Article 12", 2015)

### *6.8.3 Action for Climate Empowerment (ACE)*

The implementation of Articles 6 and 12 is officially recognized by the UNFCCC by work done as Action for Climate Empowerment (ACE). The goal of ACE is to "empower all members of society to engage in climate action, through education, training, public awareness, public participation, public access to information, and international cooperation on these issues" (*What is Action*, 2022). Within this framework is a forum specifically for youth, but again, taking place outside of the COP conference and usually in the month of July. To provide some measure of inclusion not only at COP, but in the coveted Blue Zone, ACE also hosts youth events here such as *Young and Future Generations Day* and the *Global Youth Video Competition* (2015–2019).

The Global Youth Video Competition was an interesting idea that tied in the UNFCCC's Youth Climate Report to encourage citizen (broadcast) journalism among the global community of youth. UNFCCC officials working with a UK charity called Television for the Environment (TVE) would issue a call for participants each year with two to three themes for stories, themes that were to be highlighted at that year's COP conference. TVE, a UK-registered charity, WWF-UK, and Central Television (now part of ITV) would collect the video entries, judge them, and present them in the Blue Zone on Young and Future Generations Day, previously known as World Youth Day. The top 20 films in each category or theme would be uploaded to the UNFCCC's Youth Climate Report's GIS map database and used by delegates and negotiators as a data delivery system in the Blue Zone throughout the conference. Today, there are more video competitions for youth supported by other UN agencies as well as by external organizations who have partnered with UN agencies which we will examine at length in Chap. 7.

## 6.9 Conclusion

Mutual communication through video establishes a dialogue between youth and policymakers, and one that both parties seem to have embraced, but while it still falls short of face-to-face engagement as prescribed by the Direct Approach, it does open the door to where youth want to be and where policymakers are cautiously exploring.

The next and final chapter will explore the digital dialogues taking place through this multimedia engagement more fully and will examine a framework that encourages interconnectivity between the global community of youth and international environmental policymakers demonstrating how this relationship is developing between youth and the United Nations today at the COP conferences.

## Works Cited

"About The Intercept", *The Intercept*, n.d. Accessed: December 10, 2021. Link: https://theintercept.com/about

"Article 6", *United Nations Framework Convention on Climate Change*. UNFCCC.int, 1992. Accessed: May 9, 2022. Link: https://unfccc.int/resource/docs/convkp/conveng.pdf

"Article 12", *United Nations Framework Convention on Climate Change*. UNFCCC.int, 2015. Accessed: May 9, 2022. https://unfccc.int/sites/default/files/english_paris_agreement.pdf

"Canada's Youth Policy", *Canada.ca*. Accessed: May 7, 2022. Link: https://www.canada.ca/en/youth/programs/policy.html

"Conference of Youth (COY)", *United Nations Climate Change*, n.d. Accessed: December 10, 2021. Link: https://unfccc.int/topics/education-and-youth/youngo/coy

"Global Youth Statement", *United Nations Climate Change/YOUNGO*, Glasgow: 2021. Accessed: December 10, 2021. Link: https://ukcoy16.org/wp-content/uploads/2021/10/Global-Youth-Statement.pdf

"History", *YOUNGO*. Accessed: May 4, 2022. Link: http://www.youngo.uno/about/history

"How to Organize a Phone Zap", *Incarcerated Workers Organizing Committee*, August 11, 2018. Accessed: June 1, 2022. Link: https://incarceratedworkers.org/resources/how-organize-phone-zap

"Local Youth are Leading Change", Sustainable Orillia, February 26, 2022. Accessed: June 1, 2022. Link: https://sustainableorillia.ca/local-youth-are-leading-change

"Our Impact", *youthclimatelab.org*. Accessed: May 7, 2022. Link: https://www.youthclimatelab.org/impact

"Policy Documents", *Teach the Future*. Accessed: May 20, 2022. Link: https://www.teachthefuture.uk/policy

"Press Release", Save the Children, *savethechildren.ca*. Toronto: November 5, 2021,. Accessed: December 10, 2021. Link: https://www.savethechildren.ca/article/youth-day-at-cop26-welcomed-but-not-nearly-enough

"What is Action for Climate Empowerment?", *unfccc.int*. Accessed: May 9, 2022. Link: https://unfccc.int/topics/education-youth/the-big-picture/what-is-action-for-climate-empowerment#:~:text=The%20Big%20Picture-,What%20is%20Action%20for%20Climate%20Empowerment%3F,12%20of%20the%20Paris%20Agreement

United Nations Climate Change. "Youth Building Momentum for Climate Action", *United Nations Climate Change News*, August 12, 2021. Accessed: December7,2021.Link:https://unfccc.int/news/youth-building-momentum-for-climate-action

"Who We Are", *Sustainable Orillia*. Accessed: May 18, 2022. Link: https://sustainableorillia.ca/who-we-are

"YOUNGO's Principles", *youngoclimate.org*, August 2, 2021. Accessed: May 7, 2022. Link: https://docs.google.com/document/d/1PRk4PYmcVoE37ZTKoW-0HAPTMTRIahLy3qTTeSqRI9o

"Youth Claim Their Space in Policy Circles on Climate Change", *irdc.ca*. Accessed: May7,2022.Link:https://www.idrc.ca/en/research-in-action/youth-claim-their-space-policy-circles-climate-change

Bannon, Laura. "COP26 – Will It Make a Difference?", *Inconvenient Questions International*. December 4, 2021. Accessed: December 6, 2021. Link: https://youtu.be/U8qAVjwa1h8

Brown, Alleen. "How the Fossil Fuel Industry is Attempting the Buy the Global Youth Movement", *The Intercept*, December 13, 2019. Accessed: December 10, 2021. Link: https://theintercept.com/2019/12/13/youth-climate-movement-fossil-fuel-industry/?fbclid=IwAR1aMysAy9mICW0XZGij72Gzz-y0L_sdMONLzwKeUXy3BQ3yF_rlJ6l9j8E

Burgess, Jazmin. *Climate Change: Children's Challenge*. UNICEF UK, 2013. Accessed: December 7, 2021. Link: https://downloads.unicef.org.uk/wp-content/uploads/2013/09/unicef-climate-change-report-2013.pdf

Duffy, Valerie, Gilmartin, Leo, Molay, Valery, and O'Brien, Dermott. "Climate Revolution", *youth.ie*. National Youth Council of Ireland, 2019. Accessed: May 7, 2022. Link: https://www.youth.ie/wp-content/uploads/2019/11/OWW_2019_Climate_Revolution.pdf

Gavrieli, Y., Begin, M. K., Hanin, D., Hern, M., Levitan, D., Levy, R., & Varnia, C. (2009). *Environmental Scientists and Environmental Policy Makers: Discourse Assessment and Action Recommendations*.

James W. Stoutenborough, Robert Nicholas Fette, Arnold Vedlitz, Carol L. Goldsmith, "Understanding the Communication of Climate Change Risk: Climate Scientists' Perspectives of Media Sources and Policy Makers", *Risk, Hazards, & Crisis in Public Policy* March 29, 2015. Link: https://doi.org/10.1002/rhc3.12066

Lynda Dunlop, Elizabeth Rushton, Lucy Atkinson, Eef Cornelissen, Jelle De Schrijver, Tetiana Stadnyk, Joshua Stubbs, Chrissy Su, Maria Turkenburg-van Diepen, Fernanda Veneu, Celena Blake, Saul Calvert, Clémentine Dècle, Kirndeep Dhassi, Rosalind Edwards, Greta Malaj, Jovana Mirjanić, William Saunders, Yara Sinkovec, Suzan Vellekoop & Xinyue Yuan (2022) "Youth Co-authorship as Public Engagement with Geoengineering", *International Journal of Science Education*, Part B, 12:1, 60–74, https://doi.org/10.1080/21548455.2022.2027043.

Marquardt, Jens. "Fridays for Future's Disruptive Potential: An Inconvenient Youth Between Moderate and Radical Ideas", *Frontiers in Communication*, 22 July 2020. P. 1. Accessed: December 7, 2021. Link: https://doi.org/10.3389/fcomm.2020.00048

Nichols, Bill. *Introduction to Documentary*, 3rd ed. Bloomington, IN: University of Indiana Press, 2017.

Kos, Philippa. Electronic Interview. Conducted by Mark Terry over Zoom platform. May 19, 2022.

Wynes, S., Kotcher, J. & Donner, S.D. "Can Citizen Pressure Influence Politicians' Communication about Climate Change? Results from a Field Experiment", *Climatic Change* 168, 6 (2021). Accessed: December 7, 2021. Link: https://doi.org/10.1007/s10584-021-03215-9

Wieclawek, Blythe. Electronic Interview. Conducted by Mark Terry. June 5, 2022.

Xu, Weiqi. Electronic Interview. Conducted by Mark Terry. June 1, 2022.

CHAPTER 7

# Lights! Camera! *Action*! The Use of Film by Youth and the UNFCCC

In the previous chapters of the second half of this book we have examined how opportunities to participate in global climate policy sessions organized by national governments, NGOs, and the United Nations have afforded youth new and emerging access to processes that create progressive new climate legislation. These opportunities are still not at the UN level where youth feel fully empowered and consulted as a marginalized community, but the access to present at the COP conferences and to hold a pre-COP conference of their own (COY) demonstrates an acknowledgement of the importance of youth input in these discussions. The UNFCCC, primarily through its official constituency YOUNGO, has been making strides to involve youth in the COPs in a meaningful way, short of the participatory policy action we have identified as being ideal. In her analysis of youth participation at COP22, Larissa Kwiatkowski points out the various activities and access granted to youth by the UNFCCC at COP conferences, but also recognizes their limitations.

> ...youths accredited with Observer Badges are mainly denied access to those closed negotiations, limiting the ability to allege lobbying and policy activities... These forms of participation include official Youth Delegates, YOUNGO Plenary interventions and some forms of YOUNGO lobbying done by experienced youths as they have direct access to the delegations and the decision-making. (They also participate in) most of the YOUNGO con-


M. Terry, *Speaking Youth to Power*, Palgrave Studies in Media and Environmental Communication,
https://doi.org/10.1007/978-3-031-14298-7_7

> stituency's work, such as lobbying, working in policy working groups, organising press conferences or meeting with county delegations. Through discrete lobbying and influencing behind the scenes, the aim of these activities is to advocate for youth policy suggestions. (Kwiatkowski, 32)

## 7.1 Access to the Blue Zone

Due to the restrictions UNFCCC has imposed on all non-state actors in the Blue Zone, the kind of policy participation youth demand will always fall short until these restrictions are lifted. Youth access to this important policymaking area of the COP conference is a good first step even if their activities are limited to lobbying, meetings, and influencing.

In the first half of this book, we examined how the Groundswell Approach has been able to expand to unprecedented global numbers with the use of digital communications. Never before have issues of importance to youth been able to be voiced with such amplitude and on such a global scale. These communications platforms and devices are not only relatively easy to operate for this generation of youth who grew up with this technology, but they are also quite accessible for large numbers of youth in global urban centres. We can draw a line between the UN's programs of participation and this preferred method of communications for the global community of youth to see a pattern in their mutual methodologies.

While youth are restricted to the COY conference that takes place in the days leading to the COP conference each year, and to the non-policy-creation Green Zone during the COP conferences, and to what some refer to as a tokenistic presence in the Blue Zone delivering presentations and participating in panels, their voices are nonetheless reaching the policy-makers and being used in negotiation sessions during these conferences. They are achieving a "seat at the table" by proxy through the medium of film. All non-state actors who have accreditation to the Blue Zone are subject to strict rules that limit their participation at this level. Even their badges make it clear that they are not "Participants", but just "Observers", the official designation of non-state actors in the Blue Zone. However, the UNFCCC has made it clear since 2009 that youth participation in some form is a goal they endeavour to achieve. The invitation to youth in 2009 included their contribution of video reports of climate action, research, impacts, and proposed solutions. But how did the medium of film get introduced to the Blue Zone and why was it considered a reliable surrogate to in-person presentations and peer-reviewed reports?

## 7.2 Film at COP15: *The Antarctica Challenge: A Global Warning*

At the COP15 conference held in 2009 in Copenhagen, expectations were high for significant global policy that would put the end of climate change in sight. The conference fell short of that goal, but was significant for introducing film to the Blue Zone as a data delivery system in an altogether new way. The year of 2009 marked the end of a two-year period of scientific research known as International Polar Year, a call to the world's environmental scientists to conduct concentrated research in the Arctic and Antarctica. There was eager anticipation for International Polar Year (IPY) reports at the Copenhagen conference but since many of the scientists were still embedded in the polar regions completing their research and reports, the UNFCCC began to look elsewhere for data.

Earlier that year, I had completed a documentary feature film on IPY research in Antarctica called *The Antarctica Challenge: A Global Warning*, a showcase of climate research and impacts in Antarctica featuring the work of polar scientists from the USA, the UK, Canada, and Ukraine. My production office received a call from the Communications department of the UNFCCC inquiring after the film and requesting a screening copy for review. A link to the film online was sent and 48 hours later the film was officially invited to screen in the Blue Zone throughout the two-week conference. I was also invited to attend to introduce the film and to conduct question-and-answer sessions and panels following each screening. The UNFCCC had indicated that my journalistic approach showing both the positive (yes, there are some) and negative impacts of climate change convinced them that this was a responsible, and not an incendiary, report of scientific research.

The United Nations is careful not to champion overly emotional, extreme, clearly activist documentary films, even if their messages are factually accurate and represent necessary data in the communications process between scientists and policymakers. As ecocinema theory and practice were emerging in the late 1990s, the United Nations Educational, Scientific and Cultural Organization—more commonly known as UNESCO—issued a manifesto of sorts as a guideline for those engaged in communications endeavours related to urgent global environmental issues. The document titled *Educating for a Sustainable Future: A Transdisciplinary Vision for Concerted Action* was released in November 1997. The report eschews all forms of "alarmism" in communications products such as documentary film citing that "(e)xtreme positions, while they may be useful in catching the public's attention and alerting it to pending dangers, make it difficult to move from declarations and debate into action" (Subsection 53, Section II). Instead, the reports recommend that:

> (b)oth the messages and the messengers have to appear credible and responsible. Nothing is to be gained by scaring people. Alarmist predictions that make it seem that the world is about to end are evidently not conducive to the long-term planning and action that sustainable development requires. On the contrary, it is far more effective to present problems as manageable through responsible conduct and, wherever possible, put forward a realistic solution and a means to take preventive action. (*Educating for a Sustainable Future*, 20)

Alarmist predictions that make it seem that the world is about to end are evidently not conducive to the long-term planning and action that sustainable development requires. On the contrary, it is far more effective to present problems as manageable through responsible conduct and, wherever possible, put forward a realistic solution and a means to take preventive action (Subsection 54, Section II). Since *The Antarctica Challenge: A Global Warning* met this criteria and satisfied the report's mandate for responsible communication, the United Nations Framework Convention on Climate Change made the bold proposal of including it as part of the conference programme, a data delivery system that would be accessed by DVD, online links, and on-site screenings to delegates, negotiators, and world leaders throughout the two-week conference. The film was screened 25 times in 12 days to these policy creators and hundreds of DVD copies of the film were distributed to accredited delegates who could not attend any of the scheduled screenings (Terry, *The Geo-Doc*, xiii). The film was distinguished as "the only film invited by the United Nations to screen at the 2009 COP15 conference" (*Sioux Falls Scientists*, 2009).

In this instance, the filmmaker served as a witness, an informant, and a participant in the policy-creation process. Presenting the documentary directly to the changemaker sparked policy meetings and writing sessions as new data, now in a visible form rather than merely a written, textual form, provided an additional context yielding a fuller understanding of the profiled issues (Terry, *The Geo-Doc*, xiii-xiv). Nick Nuttall, head of Communications for the United Nations Environment Programme

(UNEP) at the time, remarked that "film provides an additional context that helps deliver a fuller understanding to the policymaker".[1] The reason for this privileged opportunity for a commercially produced documentary is provided by the then-Under-Secretary General of the United Nations and Executive Director of the United Nations Environment Programme (UNEP) at the time, Achim Steiner:

> Of all the canaries in the climate coal mine, the polar regions and the mountain glaciers are singing the hardest and the loudest. *The Antarctica Challenge: A Global Warning* underlines these realities with some of the latest and increasingly sobering scientific findings. (Canadian Press/CTV News 2009)

## 7.3 Film at COP16: *The Polar Explorer*

The success of using film as a data delivery system enthused the UNFCCC. As a result, I was invited to return to the following conference—COP16 in Cancun—with another similar documentary, this time on climate research in the Arctic. The film I made was called *The Polar Explorer* and represented the first crossing of the Northwest Passage documented on film. Like its predecessor, the film was screened 25 times throughout the two-week conference and was used in negotiation sessions with the filmmaker present as a surrogate for the scientists profiled in the film.

Achim Steiner again issued a press release this time dated November 22, 2010 and entitled *UNEP Partners with Film Company for Climate Change Conference* attributing *The Polar Explorer* to enhanced policy discussions:

> The speed at which the polar regions are melting needs to be reflected in the speed with which nations come to agree on a decisive and definitive new global climate agreement. While the science is speaking loudly, it is often difficult for millions if not billions of people to witness this with their own eyes. This is the value of the film The Polar Explorer," Mr. Steiner added, "seen through the eyes of the scientists on the front-line, it brings the climate impacts at the poles to audiences in the conference halls of Cancun and the computers of the global public in order to raise the alarm but also the imperative to act. (UNEP, November 22, 2010)

[1] As told to the author December 15, 2009, COP15, Copenhagen.

The active participation the film and its filmmaker took at COP16 resulted in the creation of a resolution that was added to the Cancun Adaptation Agreement in 2010. The section of the agreement that the film informed was known as *Enhanced Action on Adaptation, Section II: Subsection 25* and marked the first time data delivered in a film format being used in the creation of a climate change policy by the United Nations:

> The Conference of Parties) (r)ecognizes the need to strengthen international cooperation and expertise in order to understand and reduce loss and damage associated with the adverse effects of climate change, including impacts related to extreme weather events and slow onset events, (i)ncluding sea level rise, increasing temperatures, ocean acidification, glacial retreat, and related impacts. ("UNFCCC" 2010)

As a result of the success of film at this level, together with the UNFCCC's new mandate of increasing the participation of youth at these conferences, a request was made by the UNFCCC's Communication department to conceive of a film project that would see youth produce video reports to be screened at subsequent COP conferences. That program would be the Youth Climate Report which we will examine further later in this chapter. The challenge was to maintain a level of professionalism delegates have come to expect and have the science explained in a more accessible way by researchers profiled in these films.

As mentioned previously in this book, it is important to remember that policymakers are rarely scientists and understanding dense, academically authored, and peer-reviewed research papers are often a challenge for those inexperienced in such disciplines. Film provides a visual context as well as the visible evidence of climate research and impacts that assist in the understanding of the data. Scholarship that supports the specific theory that the relationship between policymaker and filmmaker provides fuller understandings of scientific research can be found in the scholarship available during this same transitional year of 2009.

A study on the influence of the documentary film among government "decision-makers and policymakers" was conducted by George Ferreira, then-Programme Lead at the Ontario Ministry of Agriculture, Food and Rural Affairs in Canada. The research involved a series of interviews with his fellow bureaucrats, and in it some definitive attitudes on film as a preferred and effective information source for policymakers are revealed. In a series of films made by Ferreira and Indigenous members of the Keewaytinook-Okimakanak communities in Northwestern Ontario between 2003 and 2005, a message of financial need is delivered directly to the policymaker. Here is a typical comment from a government bureaucrat who served as a research subject in the study:

> I read significant documents all the time, but it goes out of my mind as I read the next significant thing. I keep shifting several times a day … and I think the video has much more lasting power. The video in my mind has more lasting impact than a report. Maybe it's the concentration factor as well. … It'll take me three hours to make my way through a 50-page report … but a video can take you through the concept in 15 minutes and add a human dimension that the report cannot. (Ferreira et al., 33)

This assessment reveals the trust placed on the content and messages delivered in these films by government officials. The quoted bureaucrat articulates that the visible evidence presented in film provides a credibility to policymakers that the community representing the policy to be changed is represented directly in the film, not as a mere interview subject, but as an active participant, a co-producer of the film to a degree, so that an understanding beyond what is traditionally available through text alone can be achieved. To show first-hand the conditions in which a community is living, presented by members of that community in their own words, provides more credibility of the issue being profiled. As one federal programme officer said in Dr. Ferreira's research:

> When a bureaucrat visits a (remote) community they come with their own pre-conceived notions and ideas and often, because they don't stay very long or they only see the surface, the broken down trucks all over the place, they don't get the whole story. What these videos allow the communities to do is construct the image that they see of themselves and, in a way it's actually much more honest. (Ferreira et al., 170)

The success of this participatory mode of documentary filmmaking inspired a similar approach to the production of the Youth Climate Report. The global community of youth is asked to record researchers from the global community of science to have the complicated data of climate change explained in a more accessible way, a way that the less scientifically knowledgeable policymakers of the United Nations find easier to understand. Additionally, the profiled science community is asked to showcase its findings and research methodologies to student reporters ensuring that the most significant data are highlighted, according to the world's foremost experts in the field. And finally, the global policymaking community of the United Nations can suggest presentation formats, research themes, and other content they deem necessary to assist them in specific policy creation meetings. The result is a collaborative and symbiotic enterprise involving the participation of the global communities of youth as filmmakers, science as interview subjects, and policy as changemakers (Terry, 180).

## 7.4 The Youth Climate Report (YCR)

The same approach can be applied to the global community of youth and after establishing the reliability of the content in film at the COP15 conference in Copenhagen and the following year at the COP16 conference in Cancun, the UNFCCC extended the invitation to youth to participate through documentary film reports. The reliability of the medium and the preference shown by delegates to be presented data this way made film a regular form of communications between the global scientific community and the UN policymaking community. It also provided a unique solution to the 2009 mandate to increase the level of involvement of youth in these UNFCCC conferences. The concept at the time was to curate filmed reports produced by youth from around the world with a suggested running time of three to five minutes in length. The collected films would then be stitched together in the form of a 45- to 60- min feature film to be screened at subsequent COP conferences. The structure was formalized in a program conceived by myself and Toronto environmental advocate John Kelly as the *Youth Climate Report* ("About", 2022). Its first feature film was screened in 2011 at the COP17 conference in Durban, South Africa. In total, six YCR feature films were made for the five COP conferences from 2011 and 2015 inclusively, as well as one for the United Nations Conference on Sustainable Development—known as Rio+20 in 2012. While the films were eagerly anticipated and well attended at these UN events, the consistent area of improvement suggested was for additional content. Only five to ten short films were included in each feature presentation and the UN delegates wanted more.

To answer this call, I made it a focus of my PhD research to find a way of accommodating this demand and I found the answer in the database documentary. This form of a documentary film project showcases several film units on a similar topic with a consistent narrative all in one interactive, digital space. The idea is to have several films, all variations on a theme, available to the viewer according to their selection. This multilinear approach effectively makes the viewer the editor and, to some degree, the director, as the database of film units is viewed not in the order prescribed by a linear documentary film's director and editor, but by the viewer, or in this case, the "user", themselves. Once this form of documentary was selected, a representative platform was needed to house the films. For this, I turned to Geographic Information Systems (GIS) technology. Using an experimental prototype of GIS software introduced by Google called Fusion Tables, the first multilinear, database version of the Youth Climate Report was created. It was introduced at the Paris climate summit, COP21, in 2015 (*see* Fig. 7.1). The Communications department of the UNFCCC extended the reach of the project to all participants by making it available on a series of interactive monitors scattered throughout the convention centre. They were equipped with headphones so users could engage in the project and watch their selected videos without interfering with the user next to them. Instead of presenting 10 films, we were now able to present 50 through this new platform and delegates no longer had to be available for scheduled screenings. They could engage in the project any time by using the provided monitors in the convention centre or by accessing the project on their laptops through a link provided by the UNFCCC at their convenience.

### 7.4.1 *The Geo-Doc*

Nick Nuttall, now head of Communications for the United Nations Climate Change secretariat in 2015, issued a press release praising the conference's new communications tool: "The many videos we received from around the world, placed … on this global, easy to use map, are testimony to (youth's) commitment, and I'd encourage everyone to take a look at (the) numerous examples of climate action on the ground (that) will be showcased at the upcoming UN Climate Change Conference in Marrakech in November, and the videos help underscore the fact that governments, cities, regions, businesses and investors are all stepping up to plate to jointly tackle climate change and create a greener, safer and more sustainable future for themselves and the world" ("See Inspiring Climate Action," 2016). This remediation of the documentary film, now

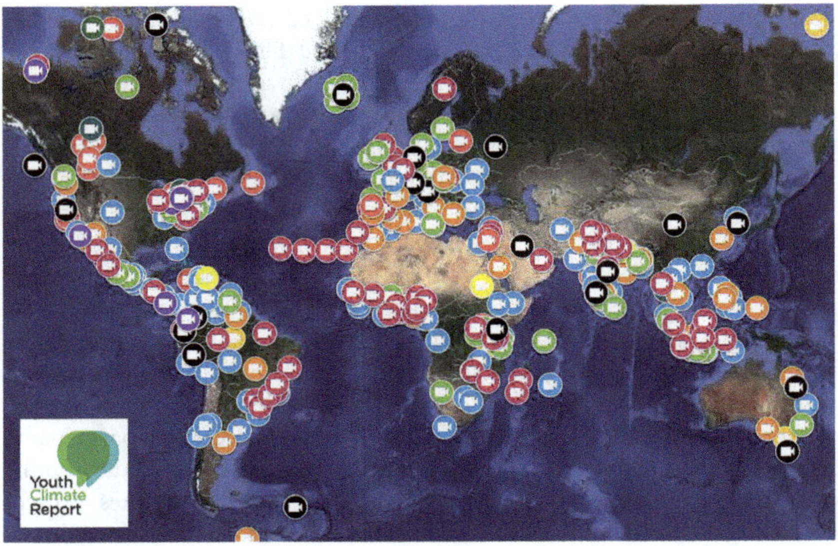


**Fig. 7.1** The Youth Climate Report, a Geo-Doc project of the UNFCCC, 2015 to present

known as a Geo-Doc,[2] seemed to answer the call of the UNFCCC for more content and better access. There was still some refinement to be made, however. The UNFCCC requested more "layers" or metadata related to each film unit on the map. In particular, links to profiles of featured scientists and their research were requested to supplement the data presented in the films.

The press release was followed by a press conference at the next climate summit, COP22, in Marrakech, Morocco. The 2016 version of YCR, now known as the Youth Climate Report GIS Project, incorporated the changes requested by the UNFCCC and was populated by 181 new video reports curated by a UNFCCC-sponsored event called the Global Youth Video Competition (GYVC), an international call to youth to create documentary short films for the COP conferences and to increase content in the YCR database. It was at a press conference that it was announced that this experimental ecocinematic Geo-Doc project was now an official partner

[2] "Geo-Doc" is a term I coined in 2019 to describe a database documentary film project presented on a platform of a Geographic Information System map.

programme of the UN Climate Change secretariat adopted under its Article 6 mandate for education and outreach (*Lights, Camera, Marrakech*, 2016). In a 2018 interview with Erik Solheim, Executive Director of the United Nations Environment Programme at the time, he reminds us that policymakers are not scientists and that data visualization projects such as YCR serve a valuable role in communicating scientific data to them. “It comes down to a need to find the best methods for communicating science. The most important thing to remember is that the decision-makers are not scientists. Certainly, they have a grasp of the basic concepts, but politicians and civil servants are also generalists. Data visualisation in (the Youth Climate Report) is one of the incredible tools available to get the science and the data across” (“Interview with Erik Solheim”, 2018).

Ecocinema documentary film theorist Helen Hughes echoes the analysis of Solheim arguing that the Youth Climate Report is a valuable communications tool for both scientists and policymakers:

> (The YCR project) has created an opportunity for young people and scientists all over the world to express their activities with respect to caring for the environment. (The project) has opened a window on the many ways in which people have been thinking about the issues. Looking at the videos from across the world there is a wonderful diversity of approach both to video making and to the kinds of science relevant to the problem of the environment and climate change research. (The films represent) models of clear, accurate, visually arresting communication… these films (show) how important such vehicles can be for the policy making process. (T)he UN online project is the most amazing visionary experiment. (“Analysis of the United Nation’s Youth Climate Report GIS Project”, 2017)

Scott MacKenzie and Anna Westerstahl Stenport analyse the Youth Climate Report and its functionality with the UN policymakers in their essay *Visualizing Climate Change in the Arctic and Beyond: Big Data in the Anthropocene, Multimodal Media Cultures, and the United Nations Conference of the Parties (COP)*:

> The UN asked Terry to produce these videos because the video format better conveys the gist of scientific data and conventional written reports to UN policy-makers, COP attendees and a global audience…The videos serve a dual purpose: they help prepare the COP delegates before the conference begins, and convey information to a global audience…Unlike many online documentary projects, Youth Climate Report deploys a highly realist aesthetic…. (MacKenzie and Stenport, 91)

## 7.5 UN Supported Video Competitions

A surprising amount of youth climate video competitions either sponsored by or supported by the United Nations have emerged in the past decade. Beginning with the Youth Climate Report in 2011 as a showcase for such youth-made documentary films, programs that assisted or recruited the production of short documentary film projects have become popular with NGOs, universities, and youth groups. Showcasing these films at the COP conferences was a significant incentive for participants and in some cases, this access served as an award when winning filmmakers were invited to attend the COP conference. While there are others, this section focuses on the five longest-running video competitions and their relationship with the UNFCCC.

### 7.5.1 *The Global Youth Video Competition*

The Geo-Doc project represents one of the first open invitations from the UNFCCC to the global community of youth to participate directly in the policymaking sessions of the COP conference. This was done by identifying specific themes for video reports through the UNFCCC's *Global Youth Video Competition* (GYVC), a program launched by UN Climate Change secretariat's Action for Climate Empowerment (ACE) and implemented by the United Nations Development Programme and the UK's Television for the Environment. Since the first competition in 2015 some of the themes provided to youth were:

- Cities and local action to combat climate change;
- Nature-based solutions for food and human health;
- Balancing use of land for people and ecosystems;
- Oceans and climate change;
- Climate friendly and resilient cities;
- Responsible production and consumption;
- Green and climate friendly jobs (Global Youth, 2022).

These themes might seem familiar as many of them relate directly to the 17 Sustainable Development Goals (SDGs) issued in 2015 at the Paris climate summit, the same year as the first Global Youth Video Competition. The synergy between the UNFCCC and the UNDP with respect to the SDGs is reflected not only by the video competition, but through another method of rewarding and incentivizing youth participation, the *SDG Action Awards.*

The objective of the program was to "highlight climate action being taken by youth through videos; providing them with a platform to identify their successes and inspire other youth and policy-makers" ("Global Youth", 2022). Films submitted to the competition were also geo-located and uploaded to the Youth Climate Report GIS Project each year. In 2021, the SDG Action Awards recognized the Youth Climate Report, with an Honourable Mention. It was explained that the awards are meant to encourage and acknowledge external initiatives aimed at connecting youth to UN climate policymakers, so organizers could not grant an award directly to YCR, but felt that its contribution to bringing youth into Blue Zone policy sessions through film could not be ignored.

This provides an excellent example of the Direct Approach and how both youth and UN are working together collaboratively despite the restrictions still in place in the Blue Zone for most non-state actors. *Access* is provided by the UNFCCC to the global community of youth to participate in the COP conference by creating these videos based on themes identified as important for each year of the COP conferences. Further *Access* is granted by inviting the winning young filmmakers to attend the Blue Zone during the COP conferences granting them access to policy meetings rarely afforded by most youth. *Respect* is given by both parties as the UNFCCC does not insist upon any style, specific content, or production methodology for the youth-made films, respecting their artistic approach and citizen journalistic responsibility for accuracy. *Respect* is given in return by youth by following the suggested themes, rules, and regulations related to the competition. *Collaboration* is achieved when the data, content, and messages provided by youth in their films are used by policymakers through the YCR database during their negotiation sessions. These films serve the purpose of informing and influencing the delegates at COP and, ultimately contribute to the content of new policy resolutions put forward at the conference as seen by the example provided in Cancun in 2010. Youth may not be sitting at the table, but their voices are being heard and their contributions are being used. There is further collaboration between the young filmmakers and the climate scientists and other experts they interview and in some cases, the young filmmakers assist the researcher in their work directly. *Finance* is also provided when the UNFCCC covers the expenses related to travel and accommodation for the winning participants. This represents one of the greatest needs for financing often identified by youth and one the UN has acknowledged in this program. And finally, *Authorship* is granted through the publication by the UNFCCC of the youth-made films bearing their names on its Youth Climate Report interactive map. Their names are not only on their films, but also on the metadata of their films when posted to the YCR database.

### 7.5.2 *The Earthbeat Challenge*

While ACE, the UNFCCC, and the UNDP all had a hand in launching and organizing the Global Youth Video Competition, films submitted to this program are not the only ones produced by youth that are showcased by the UNFCCC and YCR at the COP conferences. In 2021, a new global competition called the Earthbeat Challenge was launched by the not-for-profit organization Youth4Planet. The organization, whose slogan is "Amplifying the voices of young people through film and the media", focuses on video storytelling and offers a series of programs aimed at youth aged 14–30 to inspire and support filmmaking as a communications tool from youth to power. The Earthbeat Challenge, like the Global Youth Video Competition, provides themes for youth on which to focus their stories. In 2021, it partnered with United Nations Development Programme to focus on three themes: *Protecting and Restoring Ecosystems*, *Protecting and Restoring Land*, and *the Challenge of Climate Change*.

The video competition—the Earthbeat Challenge—is one of the activities of Earthbeat, a project that aims to enable young storytellers to share their stories and actions for a greener and more sustainable world.[3] Both project and competition are run by Youth4Planet. The Earthbeat Challenge connects with young people globally and curates their video stories on their actions and solutions to address the climate and nature emergency. The project creates opportunities for community, creativity, dialogue, and training around digital storytelling to help young people worldwide take part in shaping a better future for themselves and the planet. From its launch in August 2021, the video competition for youth has rapidly established itself as a digital project with a global reach, and as such can provide a wider voice, recognition, and inspiration to young people who are working to address the nature of climate crises at the grassroots level.

[3] https://earthbeat.youth4planet.com

Earthbeat's first video competition, on the theme of ***Restoring Balance with Nature*** attracted 77 video entries from 31 countries. From these entries ten films were short-listed each of three categories on climate, ecosystems, and land. This project partnered with UN Climate Change, the United Nations Convention to Combat Desertification, Convention on Biological Diversity, and the United Nations Development Programme and its Global Environment Facility (GEF) Small Grants Programme and culminated with events at COP26. The project received additional support from the Bertha Foundation and the Luxembourg government. The Earthbeat Challenge has built an online community attracting 1300 members to its YouTube channel, 73,000 video views, a growing following through other social media, as well as new partnerships with key international youth environmental networks such as the Youth Climate Report. Leading up to the COP the project had a team travel from Luxembourg to Glasgow via e-cycle, visiting schools and towns to engage young people around the themes and run filmmaking workshops. At the COP conference, Earthbeat joined youth activities from across the world in discussing the role of young people. Earthbeat organised a three-hour event at the Glasgow IMAX cinema in the COP26 Green Zone, which showcased climate-related films about and by young people, including the winning films from the Challenge.

Nick Turner, one of the organizers of the Earthbeat Challenge, and a previous organizer of the Global Youth Video Competition, explains that the program is designed to provide the global community of youth with an opportunity to make films about environmental issues as a means of empowering them and to assist youth in more direct engagement with policymakers, particularly at the UN level.

> One important reason that we have set up Earthbeat Challenge is that many people around the world now have a smartphone with which it is possible to record and edit video. Nearly every person with a smartphone has the potential to be a video story teller if they choose to be. The second condition is that a great many more people around the world have some access to the internet and therefore the means to share their video stories. We want to encourage young people to grasp these opportunities to make videos of local, grassroots action on climate and nature and to share them as widely as possible. From this basic concept and set of conditions Earthbeat has the following objectives:

- to support young people's capacity to have a voice on environmental issues and a platform on which to share their stories, promoting peer-to-peer dialogue on solutions via online communities;
- to catalyse relevant, youth-oriented content about key environmental issues, local solutions and green opportunities, for multiple media platforms - including YouTube and other social media;
- to empower young people to make best use of digital media storytelling and journalism to enhance their voice and increase the quality and impact of their peer-to-peer communications;
- to engage youth more directly with the debates and decision-making processes in the key international environment meetings where policymakers gather (Turner 2022).

Once again, we see a proclivity for engagement with youth and power through the medium of film as a form of artistic expression as well as a visual instrument of communication supported by various agencies of the United Nations.

### 7.5.3 *The Worldwide Youth Multimedia Competition*

Another UN-based video competition for youth is UNESCO's *Worldwide Youth Multimedia Competition* (WYMC). Unlike the Global Youth Video Competition and the Earthbeat Challenge, this contest offers a question related to the Sustainable Development Goals (SDGs) to be answered instead of themes as guidance:

> The world is not on the track to accomplish the goals in 2030, according to the SDG's 2020 report. Even before the COVID-19 pandemic, the progress had been slow. Six years after the approval of the 2030 agenda, the average SDG index of the region shows modest progress towards the goals. In terms of the SDGs, how has your region fared? According to you, what problems are there in your country that prevent the achievement of one of the SDG's goals? What real solutions could you establish for your community/country that assures success by 2030? Please do specify what goal of SDG you are focusing on. (UNESCO 2022)

The competition also opens the age range of the other UN youth-led contests and programs from 18 to 30 (12 years) to 10 to 24 (14 years). In acknowledging the disparity of the age ranges, the WYMC divides entries into three age categories: Junior (ages 10–14); Youth (ages 15–19); and

Young Adult (20–24). The competition also invites videos within the three-minute running time similar to the other competitions and the YCR recommended duration; however, unlike these UN programs, the WYMC is also open to "multimedia presentations, written essay (500–700 words), artwork, or any other manners which channels a special passion..." (UNESCO 2022). This format is similar to the citizen journalism structure and divisions of the Young Reporters for the Environment (YRE) competition that invites entries in the categories of photo essay, video report, and article. In addition to the standard certificate awarded by most youth climate competitions hosted by the UN, the WYMC offers a scholarship to UNESCO's Builders of the Universe Camp, a two-week program held at Hood College in Frederick, Maryland in the USA designed to develop leadership skills among youth. This prize is one of many offered by the WYMC to incentivize participation and it seems to be effective. The 2021 competition saw more than 5000 entries from 81 countries (UNESCO 2022).

### 7.5.4 *Young Reporters for the Environment*

The Young Reporters for the Environment competition keeps its participants engaged all year with other events such as the Spotting Nature Challenge (another contest encouraging video production) and the Global Action Days Campaign, a ten-day program featuring a series of environmental activities (some as brief as five minutes) for youth to take on and post their accomplishments on social media along with a collection of event-centric hashtags: *#GlobalActionDays, #MyActionsMatter, #GAIA 2030, #GenerationRestoration* and *#LitterLessCampaign* ("Young Reporters", 2022). These activities and social media postings build a global community of like-minded youth worldwide in yet another web-based incarnation of the Groundswell Approach. It also incentivizes them through an online presence with which they are already active and familiar. As well, the annual YRE competition awards "diplomas" to winning entries. The 2021 competition received 229 entries from 35 countries.

The Earthbeat Challenge awarded its 2021 winning entries with a screening at the IMAX Glasgow theatre during COP26. As a new competition, its numbers were relatively low, bit still impressive considering a general lack of production activity during the COVID-19 pandemic: 62 entries were submitted from 28 countries ("Earthbeat", 2022).

### 7.5.5 *The #NextGen Video Challenge*

Another youth climate video competition not directly connected to the UN, but a feeder program to the Youth Climate Report database, is the #NextGen Video Challenge in Singapore. Organizers recruit youth aged 13–30 from schools across Asia to participate in a program that includes a workshop that instructs participants how to produce documentary short films in the UN's preferred three-minute style. Like the other UN-sponsored youth climate video competitions, this one also asks its participants to work with climate themes of particular focus to the UNFCCC at its upcoming climate summit:

- Climate impacts on urban environments;
- Nature-based solutions for sustainable management;
- Protecting and restoring ecosystems (a COP26 theme);
- Climate change and my community: how we are responding (*#NextGen*, 2022).

The competition received 25 video entries, but since many were co-produced and co-directed, the competition engaged 72 participants in these productions. Unlike most UN-sponsored youth climate video competitions, the #NextGen Video Challenge awarded cash prizes to its winners: First Prize: $1000 (US); Second Prize: $800 (US); Third Prize: $500 (US); and three Honourable Mention Prizes of $100 (US) each (*#NextGen*, 2022).

In all cases, the youth climate video competitions provide an incentive to not only reward excellence, but to recruit participants. They also provide themes that are provided by the UNFCCC based on its particular areas on climate focus each given year. These themes serve to guide the participants to produce the kinds of films that will serve to inform and influence the delegates, negotiators, and policymakers of each COP conference. They also serve to inform youth of what climate matters the UN is prioritizing each year. While only the entries in the GYVC and Earthbeat are formally showcased at the COP conference, all winning or short-listed entries in every competition is added to the Youth Climate Report GIS map. Once in the digital database, they are used throughout the two-week conference by delegates in the Blue Zone and the top films of the year are presented at YCR side events and press conferences.

With either direct involvement with the UNFCCC or indirect involvement through the Youth Climate Report, the UN is extending the all-important first step in the Direct Approach: *Access*. Films made in this competition are showcased at the COP conferences either in formal events or at side events and press conferences. All winning and short-listed films are available throughout the two-week conference by way of the Geo-Doc digital space that houses the Youth Climate Report. Accessible to COP delegates via the UNFCCC website on their laptops, tablets, or mobile devices, policymakers in the Blue Zone often refer to these short, three- to five-minute films during their negotiation sessions for an immediate update on climate research, impacts, and solutions in their countries and elsewhere. Currently, there are more than 600 videos in the Youth Climate Report database on all seven continents.

Youth might not be sitting at the table in the Blue Zone during the COP conferences, but their films are, and in some cases, their filmmakers are as well to observe, make impassioned speeches, and to participate in panel discussions. The Internet in our global culture has introduced digital film and video as a preferred means of expression for youth. They are using it to communicate not only data and research, but their frustration at political inaction. They are also using it and the discipline of documentary filmmaking to express their views, opinions, and research in journalistically reliable ways to strengthen their message, amplify their voices, and to gain traction on par with those in power to participate in the process of policymaking on the issue of climate change, a global crisis that will impact on their own lives and the lives of their future generations. The United Nations has acknowledged this community and recognized their desire to use digital multimedia as an acceptable means of communication and participation for this group of non-state actors in the annual climate summits that aim to create progressive new climate policy for the world. This relationship is constantly evolving and both sides are exploring and learning new ways of working with each other. Since film as a data delivery system at the COP conferences is relatively new, the UN has relied on its educational agency, UNESCO, and its definition of this kind of media in its policy meetings. Its 1997 document *Educating for a Sustainable Future: A Transdisciplinary Vision for Concerted Action*, as mentioned earlier in this chapter, calls for "responsible" reporting and eschews "alarmism" in the presentation of this content. Consequently, a journalistic approach to the documentary film genre is preferred. The prescribed duration of these video reports by the Communications department of the UNFCCC is recommended to be three to five minutes. This length seems to consider both the conference-attending policymakers using this media as a resource as well as the content producers, many too young to have professional documentary filmmaking experience or feature-length film production budgets. In 2010, COP16 organizers explained that scheduling a one-hour documentary film (*The Polar Explorer*) was problematic due to the time restrictions of Blue Zone side events. There was an additional challenge in scheduling screenings to accommodate the conflicting schedules of densely booked delegates during COP conferences. It was suggested that instead of—and perhaps in addition to—one-hour feature films, shorter films "between three and five minutes"[4] and produced by youth could be easier accessed by delegates and involve young people at the same time.

Taking these directives into account, workshops have begun to emerge at universities around the world to train young climate activists who wish to participate in the UN policy process using film how best to structure their documentary short film style and content. These workshops support the UN initiatives of youth involvement through the video competitions they support. Most of the competitions we have examined in this chapter call for the "three and five minute" durations. There are many such workshops and training sessions conducted by NGOs, environmental groups, and at schools, but for this chapter, we will examine five that are based in universities with formal ties to YCR as they best represent the direct connection between youth and the United Nations to advance global climate policy that is both inclusive and representation of all youth groups worldwide.

## 7.6 Workshops

In 2010, Nick Nuttall, head of Communications for the United Nations Environment Programme, and later the UNFCCC itself, suggested a preferred template for short documentary films for youth to use based on duration (3–5 min), content (conference themes), and style (various standard filmmaking approaches). To better compete for their moment in the sun at the COP conferences, various youth organizations, NGOs, and universities have been designing workshops and training sessions for youth

[4] Mentioned to the author by Nick Nuttall, then-head of Communications for the United Nations Environment Programme at COP16, Cancun, Mexico, December 7, 2010.

to teach them these prescribed approaches and styles. Since many of the films that come out of these actual and virtual classrooms end up in the Youth Climate Report database, this section will examine five international workshops that YCR either conducts or supports.

### 7.6.1 *The Planetary Health Film Lab*

Established in 2019 as a research project at the Dahdaleh Institute for Global Health Research at York University in Toronto, Canada, the Planetary Health Film Lab (PHFL)[5] set out to answer the question "How can youth use film to tell stories of planetary health issues in their communities that will have impact with decision makers?" The program attempts to answer this question through four distinct phases: participant selection, a week-long, intensive filmmaking and planetary health educational workshop, ongoing mentoring, and produced planetary health short film network posting. While the actual content of the stories told by participants in their films is left open to their respective experiences, the program guides their projects with the institute's three research themes: Global Health and Humanitarianism; Planetary Health; and Global Health Foresighting (Dahdaleh 2022).

The institute defines its research themes as follows:

*Planetary Health*
A planetary health perspective views human health as a function and emergent property of the interdependent relationship between human systems and the natural systems in which we live. Our research calls urgent attention to the health of human civilization, the state of the natural systems and ecological boundaries on which our health depends, and the relationships and dynamics between these interwoven processes.

*Global Health & Humanitarianism*
We work to improve the clinical public health effectiveness of humanitarian response with scientific research, technological innovation, an appreciation of complex adaptive systems, and critical perspectives that address the realities of contemporary humanitarian policy and practice.

[5] https://www.yorku.ca/dighr/project/planetary-health-film-lab

*Global Health Foresighting*
What is the global health we want? We ask this question in multiple ways and contexts, and we pursue innovative and transdisciplinary approaches to design more effective, just, and equitable solutions that address the health challenges of the global commons ("Research Themes", 2022).

Participants in the Planetary Health Film Lab use these themes to guide the perspectives of their stories. In its first year, the PHFL recruited participants from countries around the world through a social media campaign primarily using the networks of the Dahdaleh Institute, York University, and the Youth Climate Report. The call for participants resulted in seven young people between the ages of 18 and 30 from six different countries representing three from the Global North and four from the Global South. The geographic composition of the participants also demonstrated representation from five continents: North America, South America, Europe, Asia, and Australia. The specific countries represented were: Italy, Canada, India, Australia, Colombia, and Ecuador with two participants. Each one was brought to York University and housed at the university's on-site hotel so they could work together in a collaborative laboratory environment during the week-long program. Each morning, participants would gather at the institute for a morning session of lectures, seminars, and presentations related to planetary health research, professional filmmaking training, youth activism studies, United Nations policy and conference overviews, and communications theory. In the afternoons, they would work on their film projects in the institute's offices. At the end of the week, the finished films were presented publicly at what organizers call a *Micro Film Festival*, so named to reflect the short, three-minute runtimes of the films. The young filmmakers were then provided the opportunity to conduct a question-and-answer session with the public audience. The festival concludes with the presentation of a framed certificate of completion, an important aspect to the *Authorship* component of the Direct Approach that underscores the addition of their films to the Youth Climate Report database with their names published on the UNFCCC website as well as on their own films.

The PHFL program has four Project Leads: myself, James Orbinski, Kate Tilleczek, and Deborah MacDonald. These researchers bring a diverse background of educational disciplines and professional experience to the program ensuring a comprehensive workshop experience for the young participants. James Orbinski, a medical doctor, is also a professor and the inaugural director of York University's Dahdaleh Institute for Global Health Research. Most notably, he has worked providing medical humanitarian relief in situations of war, famine, epidemic disease, and genocide with *Médecins Sans Frontières/Doctors Without Borders* (MSF). He was elected International President of MSF from 1998–2001, and accepted the Nobel Peace Prize awarded to MSF in 1999 ("James Orbinski" 2022). Kate Tilleczek is a professor and researcher who holds the Tier 1 Canada Research Chair in *Youth, Education & Global Good* in the Faculty of Education at York University. She is also the founder of the *Young Lives Research Lab*, a group that conducts research in matters related to youth with a particular emphasis on planetary health issues. Her lab "employs global, intercultural, and interdisciplinary approaches to collaborative research *with and for* young people and their communities" ("Dr. Kate Tilleczek" 2022). Deborah MacDonald, Senior Research Associate and Manager at the Young Lives Research Lab, "contributes to the collaborative design, implementation and evaluation of a variety of multidisciplinary research projects and partnerships based in youth and community wellbeing, education, and better understandings of the impacts of the modern world on young people today" ("Deborah MacDonald" 2022). The final co-lead on the Planetary Health Film Lab is myself, a professor of environmental studies at York University and of documentary film studies at Wilfrid Laurier University in Waterloo, Canada. I also serve the UNFCCC as the Executive Director of the Youth Climate Report providing a unique access to the United Nations and its annual climate summits to participants of the Planetary Health Film Lab.

The program's first iteration received funding from Canada's Social Science and Humanities Research Council and subsequent years by the Hunter Family Foundation. The first films produced were:

- *Welcome to Venice* (Italy): An examination of the impacts of rising sea levels on Venice;
- *Ecuador* (Ecuador): An examination of how changing weather patterns are affecting Indigenous communities in Ecuador;
- *Climate Change and Anxiety* (India): An examination of mental health issues in India resulting from climate change anxiety;
- *My Australia* (Australia): An examination of the impacts of climate change on Australia with emphasis on the 2019 forest fires;

- *The Subachoque River: Pollution and Climate Change* (Colombia): An examination of how discarded plastic and climate change have affected the Subachoque River and the people who rely on it in Colombia;
- *My Relationship with the Ocean* (Canada): An examination of changes to the ecosystem of the oceans as a result of microplastics.
- *The Páramo* (Ecuador): An examination of the impacts of climate change to the ecoregion of Ecuador located above the treeline in the Andes Mountains.

All seven films were posted to the Youth Climate Report and presented at the COP25 UN climate summit in 2019 held in Madrid. During this conference delegates accessed the YCR project watching the PHFL films during the two-week conference close to 1500 times collectively.[6] The content of these films informed the delegates of the COP25 conference in both the Green and Blue Zones contributing directly to policy discussions among those delegates engaged in such activities.

This collaboration between the data collector (YCR), the citizen journalist (the participants), and the policymaker (the United Nations) establishes an ecology of communication and education which supports the respective goals of each actor. Unlike many similar workshops, one of the unique features of the Planetary Health Film Lab borrows from the concept of the Groundswell Approach to extend lessons learned in both theory and practice with their own communities following the workshop: "(P)articipants were encouraged to showcase their films when they returned to their home countries and share what they learned at the film lab with other young filmmakers to expand the reach of the project and produce more content that amplifies the voice of young people worldwide" (Terry, *The Lancet Planetary Health*, e550).

Subsequent Planetary Health Film Labs have focused on amplifying Indigenous voices in remote communities throughout the Circumpolar Arctic and in Ecuador. Young filmmakers from the Circumpolar Arctic participated in the 2021 PHFL to tell stories related to planetary health in their home communities and in the summer of 2022 the Indigenous communities in Ecuador will feature 20 young filmmakers telling their stories in their native languages of Kichwa and Schuar. As environmental issues in

[6] Based on number of views reported by youtube.com for each film during the dates of December 2 to 13, 2019, the dates of the COP25 conference.

these parts of the world are measurably different than the rest of the world, so too, are their impacts on the Indigenous people that live there. These stories become critical in the global conversation of planetary health and having these stories represented by film in a single digital space of a GIS mapping project affords the policymaker with the visible evidence necessary to assimilate the data in a more inclusive way. The interdisciplinary presentation of data in this regard provides a variety of perspectives and new data and metadata that contribute to a fuller understanding and can lead to better solutions that take into account the anthropogenic contributions to the ecosystem of the planet ("GIS and the Environmental Humanities" 2022).

### *7.6.2 Ghana Youth Video Program*

A similar program launched in 2021 and aimed at training youth to produce ecocinematic documentaries for the UNFCCC's Youth Climate Report is the *Ghana Youth Video Program*, a research project out of the History Department at Wilfrid Laurier University in Waterloo, Ontario, Canada and funded by Canada's Social Science and Humanities Research Council. This program, as its name suggests, focuses on young filmmakers and the climate impacts to various communities of Ghana. In this program, 12 youth aged between 18 and 30 participated in making documentary short films between three and five minutes. Three older Ghanaian videographers served as advisors and producers on four of the films. A one-day workshop conducted virtually in 2021 due to the pandemic, provided technical training for producing films according to the preferred style of YCR. Participants were given approximately two months to work on their projects while consulting with project co-leads Jeff Grischow, a professor of African History at Wilfrid Laurier University, Magnus Mfoafo-M'Carthy, a professor of social work, also at Wilfrid Laurier University, and myself.

The program concluded with 12 completed video reports and all of them approved and added to the UNFCCC's Youth Climate Report database:

- *The Rise in Sea Level as a Result of Climate Change in Coastal Communities in Ghana*: An examination of recent flooding resulting from rising sea levels to coastal communities.

- *The Impact of Climate Change on Rural and Agricultural Communities*: An examination of the negative impacts of climate change to Ghana's farming communities.
- *The Fight against Poseidon*: An examination of the socio-economic impacts of erosion and flooding to Ghana's coastal communities due to rising sea levels.
- *Urban Action towards Climate Change*: An examination of how the government of the capital city of Accra is responding to the impacts of climate change.
- *Our Dying Oasis*: An examination of extreme heat and drought conditions in Ghana.
- *Climate Change to Agriculture to My Community*: An examination of the impacts of climate change to Ghana's farming communities.
- *Effects of Climate Change on Rural and Agricultural Areas*: An examination of the impacts of climate change from the perspective of local farmers.
- *The Surge*: An examination of how rising sea levels and string winds are combining to flood the city of Keta in the Volta region of Ghana.
- *Green Zongo*: An examination of how climate migration is placing new stresses on urban environments.
- *Smart Farmers*: An examination of new adaptation techniques being developed by farmers to cope with climate change impacts.
- *Climate Impact in Anyamam*: An examination of the impacts of climate change in Anyamam, in the Greater Accra Region of Ghana.
- *Impacts of Climate Change on Rural and Agricultural Areas*: An examination of the negative impacts of climate change to Ghana's rural and agricultural areas (Youth Climate Report 2022).

A quick analysis of these films reveals some common themes regarding climate change impacts in Ghana, namely: rising sea levels and farming. Rising sea levels have especially impacted the southeastern cost of Ghana, where entire villages have been displaced.[7] The challenge of climate change for farming is not simply a matter of too little rain or rising temperatures. Rather, it is the unpredictability of the rainy seasons. Sometimes there is too little rain, sometimes too much, and the rainy periods have become very unpredictable.

[7] This happened in one of the filming locations a few weeks after the completion of the film. The filmmaker returned and added new footage of the flooding to his film.

These become important when party delegates in this country participate in negotiating sessions at the UN's climate summits. When other delegates consider the scientific research papers and government white papers from Ghana regarding these issues, they will be able to turn to the video reports made by Ghanaian youth for valuable visible evidence of these impacts. Here, the Youth Climate Report serves the requirements of the Direct Approach: it provides *Access* to youth to speak directly to UN policymakers using film as the medium of communication; *Respect* is shown by youth to power by following the recommended style and format of video reports preferred by the UNFCCC and *Respect* is further shown by the policymakers who seriously consider the content and messages produced by youth in these films; *Collaboration* occurs when the young filmmakers seek out government officials and scientific or academic experts for interview subjects, providing the credible source material UN policymakers have come to rely on. These reports are presented in a manner that is more accessible than complex documents that can be sometimes be a challenge for some policymakers to comprehend fully; *Finance* is provided through this program with funds being made available directly to participating youth and their Ghanaian academic organizer counterparts to facilitate their filmmaking; and finally, *Authorship* exists for the young filmmakers within the films' credits as well as in the metadata of each film's posting in the Youth Climate Report database.

### *7.6.3 Young Reporters for the Environment (YRE)*

One of the more established programs for youth to serve as citizen journalists for the UN climate summits is *Young Reporters for the Environment* (YRE) coordinated by the Foundation for Environmental Education out of the University of Copenhagen. The program was founded in 1991 when Philippe Saugier founded the Ozone Project and organized a group of youth to travel to northern Sweden and Finland to report on the work of scientists stationed there studying the Ozone Hole. This series of field visits and expert interviews attempted to fully understand the ozone issue, and to report it to the widest possible audience in real time. Germany, Denmark, Finland, France, Latvia, Norway, Poland, and Switzerland were the first to send "youth envoys", establishing an international collaboration within Europe that would soon turn global. The first formal incarnation of the program took place in Luxembourg in 1994 ("A Brief History" 2022).

Unlike the other workshops we have examined, video reports are not the only medium these young citizen journalists use in this program to communicate with the policymaker. Training is also provided for those who wish to tell their environmental stories through photo essays and articles. The filmmaking workshop is again provided by the Youth Climate Report to ensure participants' video reports adhere to the established style prevalent in most films in the database. Using this multimedia approach, YRE attracted 229 entries of articles, photos, and videos from 35 countries to its international competition in 2021 (YRE 2022).

Some of the video entries in the 2022 competition include;

- *Now That I Am Inside, I Am a Pet, Right?* (India): A report on a rescue mission of a baby squirrel conducted by the filmmaker;
- *Toilet for All* (India): A report on a student campaign to raise funds for a toilet in their school;
- *Be the Change!* (India): A report on an initiative that involves youth in programs related to SDG 6: Clean Water and Sanitation;
- *One Planet, Two Schools* (India and France): A collaborative report on plastic pollution made by the students of two schools, one in India and the other in France;
- *How to Help Out Our Planet* (Northern Ireland): A report on recycling;
- *Web Summit 2021: Startups for Sustainability* (Portugal): A report on how technology can be used by youth for a climate revolution;
- *Footbridges, An (In)Sustainable Tourism?* (Portugal): A report on the growth of footbridges for ecotourism and their impacts on the environment;
- *Nas Daily: The One-Minute Guy* (Portugal): A report on a press conference given by social media influencer Nas Daily and his take on how he trusts social media platforms like Facebook to lead the way in environmental reform more than governments;
- *Riversides: How to Generate Awareness Around Us* (Argentina): A report on the benefits of removing garbage from riverbanks;
- *Litter Less: Sustainable Consumption* (Brazil): A report on sustainable consumption in Brazil;
- *Small Actions = Great Impact* (Spain): A report on Posidonia and its importance as underwater vegetation to ocean ecosystems.

These are just some of the videos submitted to the 2022 program of Young Reporters to the Environment. All videos will be added to the Youth Climate Report in time for the COP27 conference in Egypt in November, 2022.

The goals of this program are well-defined:

**Empower**: YRE empowers young people to have their say on environmental issues in their locality which they feel are not being adequately addressed. It gives them an outlet for their frustrations and a platform from which to make a difference.

**Unite**: Coordinated by the Foundation for Environmental Education and active in 42 countries worldwide, the YRE programme brings students of all backgrounds together with a sense of common purpose.

**Develop**: YRE helps participants to develop skills and acquire knowledge which will stand to them for the rest of their lives: communication skills, individual initiative, team work, critical analysis, social responsibility, and leadership skills are all honed through involvement in the programme.

**Drive**: A competition is held annually to encourage young people from all over the world to push themselves to their limits in the search for a story which will ultimately bring to the fore a real and current local environmental issue.

Together, these goals serve the workshops that train its young participants not only to produce citizen journalist reports in various media, but to give them the practical and leadership skills to better communicate with those in power to assist them in creating progressive new environmental policy change for their generation and for those to come.

### *7.6.4 The #NextGen Video Challenge*

This Singapore-based program is aligned with World News Day organizers and provides a rare opportunity for young people to have their films presented on this international day that celebrates journalists and journalism. Organized by the World Editors Forum and The Canadian Journalism Foundation, World News Day, the event brings together more than 500 news organizations in discussions about climate change (World News Day 2022).

Organizers of the #NextGen Video Challenge work with the Youth Climate Report to present workshops and training sessions to guide young participants to produce short documentary films for this global media event as well as for the YCR database. Participants are selected through elementary and secondary schools throughout Asia. While 25 films were made in 2021, often co-produced by two or three students, three winners and three runner-ups were announced within the four themes identified previously when this chapter examined the program's competition.

All four themes reflect the areas of focus for that year's UN climate summit, COP26, in Glasgow, Scotland. All films made in the workshop were added to the YCR database providing UN delegates a much-needed depiction of climate research, impacts and solutions in Asia, a significant area with limited video representation prior to 2021. The winning films and the schools which the filmmakers attend in the program's 2021 competition were:

- *The Master Waster Chris* (Nanyang Polytechnic): An animated documentary exploring how we are wasteful in our everyday lives;
- *The Wild Side of Singapore: Coastal Protectors* (ITE Central): An examination of how urban sprawl is reducing the numbers of mangroves and forests in Singapore;
- *Unconscious Decisions* (Nanyang Polytechnic): A minimalistic comparison of two people living two different environmental lifestyles, one is wasteful and the other recycles;
- *What If We Hear from Them?* (ITE East): An interview with two children who share what they think about climate change and how they are doing their part;
- *Nature-Based Sustainable Swaps!* (Methodist Girls' School): Another animated documentary done in a videogame style suggesting natural alternatives to daily lifestyle choices;
- *Hope for Climate Change* (Methodist Girls' School): An examination of the hope youth have for a future without climate change.

All 25 films made in this workshop focused on the themes reflected by the COP26 conference and as a result, were valuable in providing delegates with the perspective of youth from Asia on these topics. The participants also received the benefit of learning journalistic documentary filmmaking skills and had their work presented at another global forum, World News Day, September 28, 2021.

## 7.7 Conclusion

Collectively, these film workshops and competitions reflect a means by which the UN is currently engaging in their attempt to participate with youth in the policy process of their annual climate summits, the COP conferences. Since no non-state actor participates directly in policy negotiations in the Blue Zone of these conferences, youth are still absent from these discussions physically, but as they are in so many other engagements in their lives, they are there digitally. These films serve a dual purpose: they offer the visible evidence of climate research, impacts, and solutions that provide a visual context to the written data provided by the scientific community that untrained policymakers sometimes struggle with; secondly, they serve to represent the voice of the global community and what stories best reflect the climate issues of most importance to them. With the UN providing themes for each conference, the conversation becomes a dialogue between the two parties. It may never replace the in-person participatory process we have seen so many local and national governments engaging in, but perhaps one day, these models will provide the framework of engagement for the youth and the UN to sit together at the table of policy negotiation and creation in the hallowed halls of the COP conference's Blue Zone.

The Direct Approach is very much addressing the key factors of engagement youth have identified as being missing from most participatory politics, factors that have sometimes prevented their involvement even when it was invited: *Finance* and *Authorship*. In some cases, the workshops and competitions offer participants financial support in either attending the workshop or the COP conference itself. Other times, cash prizes provide the means for young filmmakers to advance their activist documentary trade or to finance travel to climate-related events like the COP conferences or meetings in the capital cities of their countries so they may engage meaningfully in local and national policy discussions. *Authorship* is addressed and provided in a couple of unique ways: each film includes the names of the young filmmakers in its credits. These names are also included in the metadata of each film's posting in the UNFCCC's Youth Climate Report database giving authority and accountability to the messages and content of the films to the UN while at the same time providing the means for follow-up engagement, something the Groundswell Approach cannot provide within its mass gatherings of protest marches. Here, as loud as the collective voice is, the individual youth is nameless; through the Direct Approach, as it is applied to these UN-youth film opportunities, each one of the many are represented, heard, and recognized.

## Works Cited

"About", *Youth Climate Report*. Accessed: May 21, 2022. Link: http://youthclimatereport.org/about

"A Brief History of YRE", *Young Reporters for the Environment*. Accessed: May 27, 2022. Link: https://www.yre.global/our-history

"Analysis of the United Nation's Youth Climate Report GIS Project," *Youth Climate Report*. youthclimatereport.org. Accessed: May 31, 2022. Link: http://youthclimatereport.org/news-cat/analysis-united-nations-youth-climate-report-gis-project-helen-hughes

"Canadian Film to Be Shown During UN Climate Meeting." Canadian Press/CTV News, August 18, 2009. Web. Accessed May 20, 2022. https://www.ctvnews.ca/canadian-film-to-be-shown-during-un-climate-meeting-1.426585

*Dahdaleh Institute for Global Health Research*, yorku.ca/dighr. Toronto: York University. Accessed: May 27, 2022. Link: https://www.yorku.ca/dighr

"Dr. Kate Tilleczek", *Young Lives Research Lab*, younglivesresearch.org. Toronto: York University. Accessed: May 27, 2022. Link: https://www.younglivesresearch.org/drkate-tilleczek

"Deborah MacDonald", *Young Lives Research Lab*, younglivesresearch.org. Toronto: York University. Accessed: May 27, 2022. Link: https://www.younglivesresearch.org/deborah-macdonald

"Earthbeat Challenge, The", *youth4planet international* [sic], Luxembourg. Accessed: May 21, 2022. Link: https://earthbeat.youth4planet.com/earthbeat-challenge

*Educating for a Sustainable Future: A Transdisciplinary Vision for Concerted Action*. United Nations Educational, Scientific and Cultural Organization (UNESCO), 1997.

"James Orbinski", *Dahdaleh Institute for Global Health Research*, yorku.ca/dighr. Toronto: York University. Accessed: May 27, 2022. Link: https://www.yorku.ca/dighr

"GIS and the Environmental Humanities: How Citizen Scientists, Civil Servants, and Researchers are Teaming Up to Study and Solve Environmental Issues", by Mark Terry, *The Emerging Role of GIS in the Environmental Humanities*, ed. Mark Terry and Michael Hewson. Lanham, MD: Lexington Books, (*submitted for publication, June,* 2022).

"Global Youth Video Competition", *United Nations Climate Change*. Accessed: May 21, 2022. Link: https://unfccc.int/topics/education-and-outreach/events%2D%2Dmeetings/global-youth-video-competition

"Interview with Erik Solheim." *youthclimatereport.org*, April, 2018. Web. Accessed May 21, 2022. http://youthclimatereport.org/?p=2428&preview=true

"Lights, Camera, Marrakech." United Nations Climate Change. Marrakech, Morocco, November 9, 2016. Web. Accessed May 21, 2022. https://unfccc.int/topics/education-and-outreach/events%2D%2Dmeetings/global-youth-video-competition/2016-global-youth-video-competition-on-climate-change

*#NextGen Video Challenge*, Singapore. Accessed: May 21, 2022. Link: https://worldnewsday.org/meet-nextgen-video-challenge-winners

"Remote Aboriginal Communities in Northwestern Ontario, Canada." Guelph: *Journal of Rural and Community Development* 4, no. 2 (2009).

"Research Themes", *Dahdaleh Institute for Global Health Research*, yorku.ca/dighr. Toronto: York University. Accessed: May 27, 2022. Link: https://www.yorku.ca/dighr

"See Inspiring Climate Action." Resource: United Nations Framework Convention on Climate Change, October 19, 2016. Web. Accessed May 21, 2022. https://unfccc.int/news/see-inspiring-climate-action-entries-to-global-youth-video-competition

"The Antarctica Challenge: A Global Warning." *Sioux Falls Scientists*, n.d. Web. Accessed May 20, 2022. http://www.siouxfallsscientists.com/the-antarctica-challenge.html

*UNEP Partners with Film Company for Climate Change Conference.* United Nations Environment Programme Press Release. Cancun: United Nations Environment Programme, November 22, 2010. Accessed May 22, 2022. Link: http://youthclimatereport.org/wp-content/uploads/2022/05/UNEP-Press-Releases-November-2010-Mark-Terry.pdf

"UNESCO Clubs 2022 Worldwide Youth Multimedia Competition", *UNESCO*. Accessed: May 21, 2022. Link: https://www.unescousa.org/youth-multimedia-compe

UNFCCC. (2010). *The Cancun Agreements*, Decision 1/CP.16, Report of the Conference of the Parties on its sixteenth session, held in Cancun from 29 November to 10 December 2010 (FCCC/CP/2010/7/Add.1). Bonn: United Nations Framework Convention on Climate Change Secretariat.

*World News Day*, worldnewsday.org. Accessed: May 27, 2022. Link: https://worldnewsday.org

"Young Reporters for the Environment", Foundation for Environmental Education, Copenhagen. Accessed: May 21, 2022. Link: https://www.yre.global

"Youth Climate Report", *United Nations Framework Convention on Climate Change*. Accessed: May 27, 2022. Link: https://unfccc.int/topics/education-youth/resources/youth/youth-climate-report

"YRE International Competition", *Young Reporters for the Environment*. Accessed: May 27, 2022. Link: https://www.yre.global/yre-competition

Ferreira, George, Ramirez Ricardo, and Allan Lauzon. "Influencing Government Decision Makers through Facilitative Communication via Community-Produced Videos: The Case of Remote Aboriginal Communities in Northwestern Ontario, Canada." Guelph: *Journal of Rural and Community Development* 4, no. 2, 2009.

Kwiatkowski, Larissa. "Paths to Meaningful Youth Involvement at the International Climate Change Negotiations: Lessons from COP22 in Marrakesh." (2017).

MacKenzie, Scott, and Anna Westerstahl Stenport. "Visualizing climate change in the Arctic and beyond: Participatory media and the United Nations Conference of the Parties (COP), and interactive Indigenous Arctic media." *Journal of Environmental Media* 1.1 (2020): 79–99.

Terry, Mark. "Amplifying the voice of youth through planetary health films." *The Lancet Planetary Health* 4.12 (2020a): e549-e550.

Terry, Mark. *The Geo-Doc: Geomedia, Documentary Film, and Social Change*. Basingstoke: Palgrave Macmillan, 2020b.

Turner, Nick. Electronic interview. Conducted by Mark Terry, June 1, 2022.

# Correction to: A Seat at the Table

**Correction to:**

**Chapter 4 in: M. Terry, *Speaking Youth to Power*, Palgrave Studies in Media and Environmental Communication, https://doi.org/10.1007/978-3-031-14298-7_4**

Owing to an oversight on the part of the publisher, Sect. 4.11 was missing in the initially published version of this book. The additional text has led to a slight change of the pagination of the subsequent chapters. The publisher apologizes to the author and his readers.

The updated original version for this chapter can be found at
https://doi.org/10.1007/978-3-031-14298-7_4



M. Terry, *Speaking Youth to Power*, Palgrave Studies in Media and Environmental Communication,
https://doi.org/10.1007/978-3-031-14298-7_8

# Conclusion

This book examines two distinct yet connected approaches to youth climate activism with a focus on their impacts on informing and influencing environmental policy with the United Nations. The *Groundswell Approach* looks at historical precedents of youth activism and how these movements grew in size and sway as young people turned their attention to the global issue of climate change in a time with unprecedented digital telecommunications at their fingertips. The proposed PLAN framework is based on the successful components of the movement and expands on achievable areas of activism to maximize the goal of participatory policy engagement.

The *Direct Approach* examines a smaller movement of young people worldwide who instead of or in addition to protesting in the streets, find favourable impact in influencing policy by way of contacting policymakers directly and working with them to voice their concerns in a much different way. A comparison of the two approaches finds positive results to each: the Groundswell Approach, led by such high-profile youth climate activists such as Greta Thunberg and Autumn Peltier, yield measurable impact in terms of participant numbers, global coordination in protests, information exchange through web-based social media platforms, email, and communication software. The approach evidently accelerates public and political attention to a global crisis many politicians are reluctant to address. While this overwhelming influence spreads to conventional media sources and



M. Terry, *Speaking Youth to Power*, Palgrave Studies in Media and Environmental Communication,
https://doi.org/10.1007/978-3-031-14298-7

other non-youth groups around the world, its pressure opens the door with policymakers but not enough for meaningful entry. Governments on all levels have traditionally resisted bending to the demands of protestors who they perceive as "anti-government and anti-system" rather than pro-progressive policy (Şöhret, 20) and continue to do so today in (a lack of) response to the global youth climate protests. Youth climate activists are lobbying for significant climate legislation first and are not necessarily intent on dismantling systems or overthrowing governments, but will support such action if necessary. This perception can make government officials reluctant to engage with those who can be seen as reformists at best, anarchists at worst.

> Although global (youth climate) protests do not currently produce major political changes around the world, these events are an important signal in terms of showing the limits of changing relations within countries and between governments and citizens. Countries and governments that can understand the underlying complaints of the protests and adapt their systems to the expectations of their citizens will achieve greater political stability, internal cohesion and ultimately global competitiveness. (Şöhret, 26)

Instead, government officials on national and regional levels, and to a lesser yet growing extent at the UN level, have experimented with the Direct Approach, inviting climate policy participation from youth communities in forms that acknowledge the inexperience of youth, yet respect their position in having a say in policies that affect them today and in the future. As such, this book examines these methodologies and proposes a five-step framework to be considered and implemented in any "youth-and-power" participatory policy engagement: *Access*, *Respect*, *Collaboration*, *Finance*, and *Authorship*. This proposed framework, informed by three previous models of youth participation in policy engagement, and first-hand field research in youth activism at the UN level through 12 years of participation in the COP conferences, identifies the key aspects of historically successful participation, namely, *Access*, *Respect*, and *Collaboration*, and also recognizes a couple of new areas that have contributed to a lack of complete and/or meaningful policy participation: *Finance* and *Authorship*. When all five components are present, we see a methodology emerge that seems to satisfy both youth and UN policymakers for the time being: short documentary film reports presented at the COP conferences and available to policymakers through the UNFCCC's

Youth Climate Report database. This is far from a perfect model as in-person policy participation in the privileged Blue Zone is still missing, but through the digital surrogacy of video reports produced by youth, their voice is not compromised and a valuable component of climate comprehension is provided through the visible evidence of film. The temporal and spatial affordances of GIS technology in its Geo-Doc format also advance fuller understandings of each of the film units in the database when compared with each other through these layers. While each film unit tells its story with an explicit narrative, comparing these projects by time and space yield implicit narratives and new data for the policymakers (Terry, 146).

Of particular affinity to the youth of today, the medium of digital film is accessible and its dissemination methods familiar to many making it an ideal communication tool for production and distribution of climate research, impacts, and solutions, and one the UN has found reliable and beneficial since 2009. The next step, of course, will be consultation with representative youth communities in person and at the same table of negotiation as UN policymakers and negotiators when the UNFCCC eventually recognizes global communities of marginalized people and non-state actors such as youth, Indigenous people, the disabled, and so on, officially in a contributory capacity in Blue Zone policy sessions. Until they are ready to take that step, the global community of youth and its video workshops, competitions, and reports will represent them in their place.

## Works Cited

Şöhret, Mesut. "Root Causes of Global Protests," *Proceedings Book*. University of Vienna: International Conference on Strategic Research on Scientific Studies and Education, Omer Faruk Tufekci, editor, 2020.

Terry, Mark. *The Geo-Doc: Geomedia, Documentary Film, and Social Change*. Basingstoke: Palgrave Macmillan, 2021.

# Index[1]



[1] Note: Page numbers followed by ‘n’ refer to notes.






M. Terry, *Speaking Youth to Power*, Palgrave Studies in Media and Environmental Communication,
https://doi.org/10.1007/978-3-031-14298-7

Printed in the USA
CPSIA information can be obtained
at www.ICGtesting.com
CBHW061319080524
8244CB00004B/152

9 783031 142970